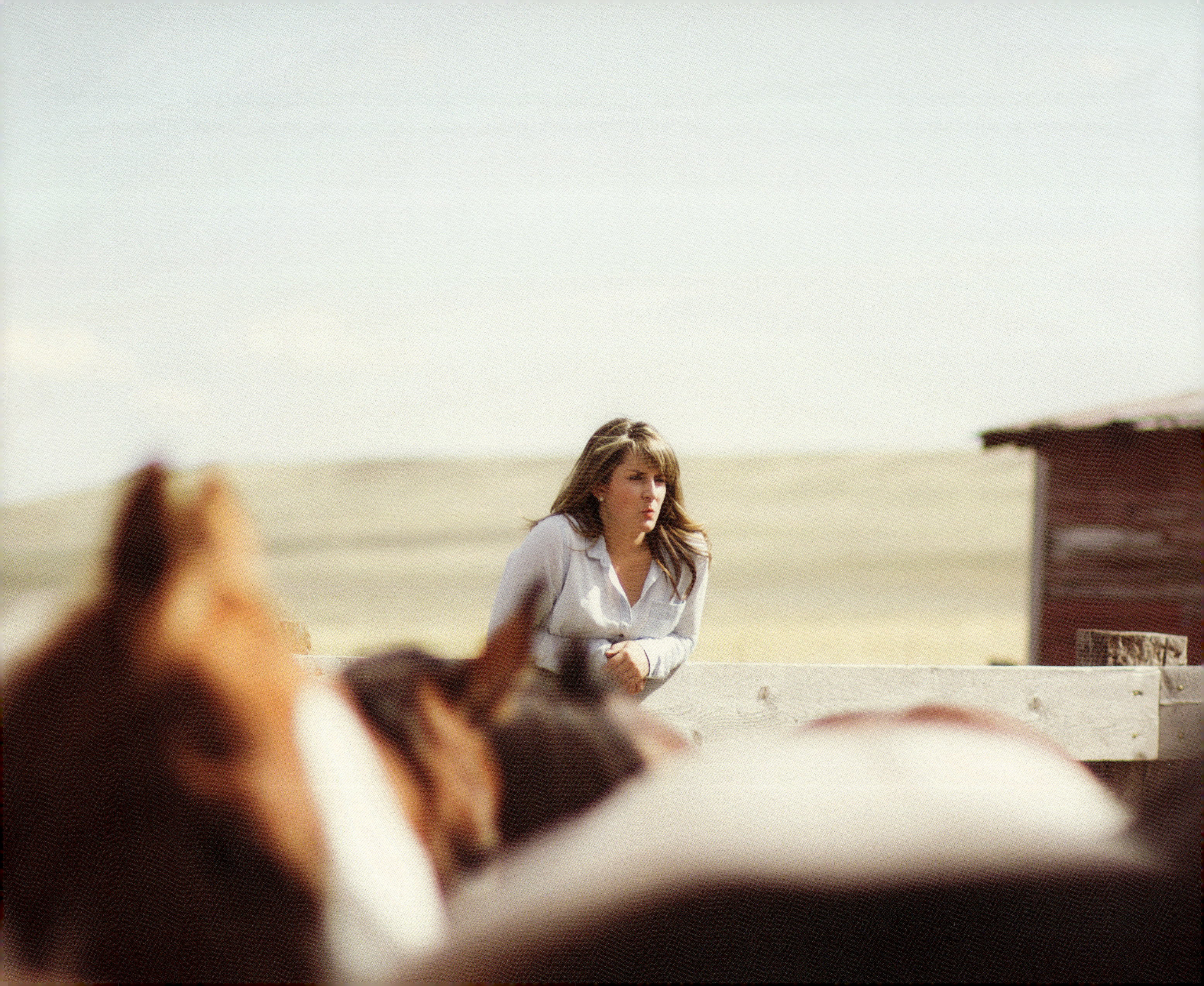

FARMHER®

SHINING THE LIGHT ON WOMEN IN AGRICULTURE

Photographs by Marji Guyler-Alaniz

Iowa Public Television
Iptv.org

Friends Iowa Public Television FOUNDATION

Copyright © 2016 Iowa Public Television and Friends of Iowa Public Television Foundation. All rights reserved. Reproduction of the whole or part of the contents without written permission from publisher is prohibited. Designed by Phong Duong, Alisa Dodge and Joe Bustad.

ISBN 978-0-692-69990-4

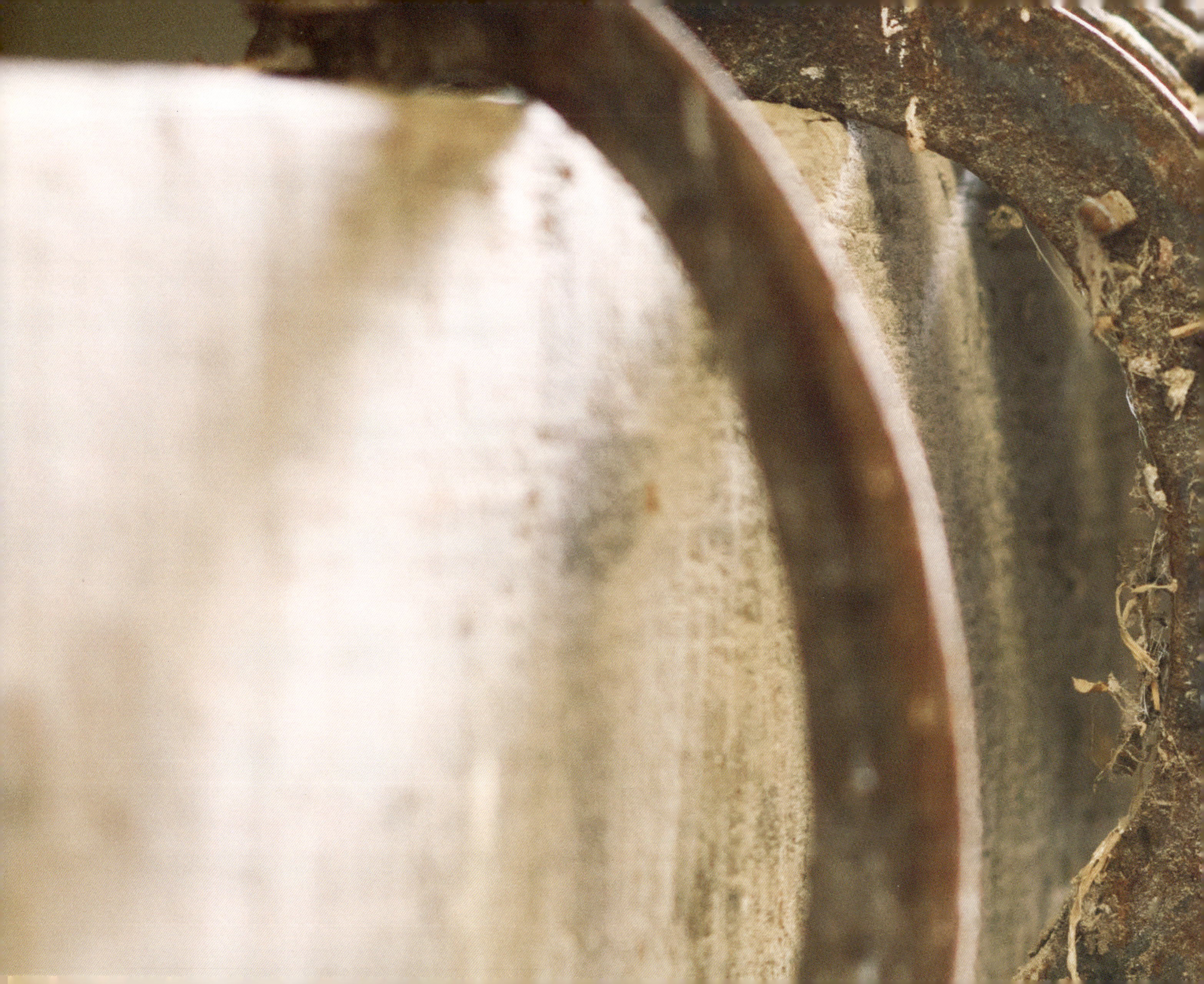

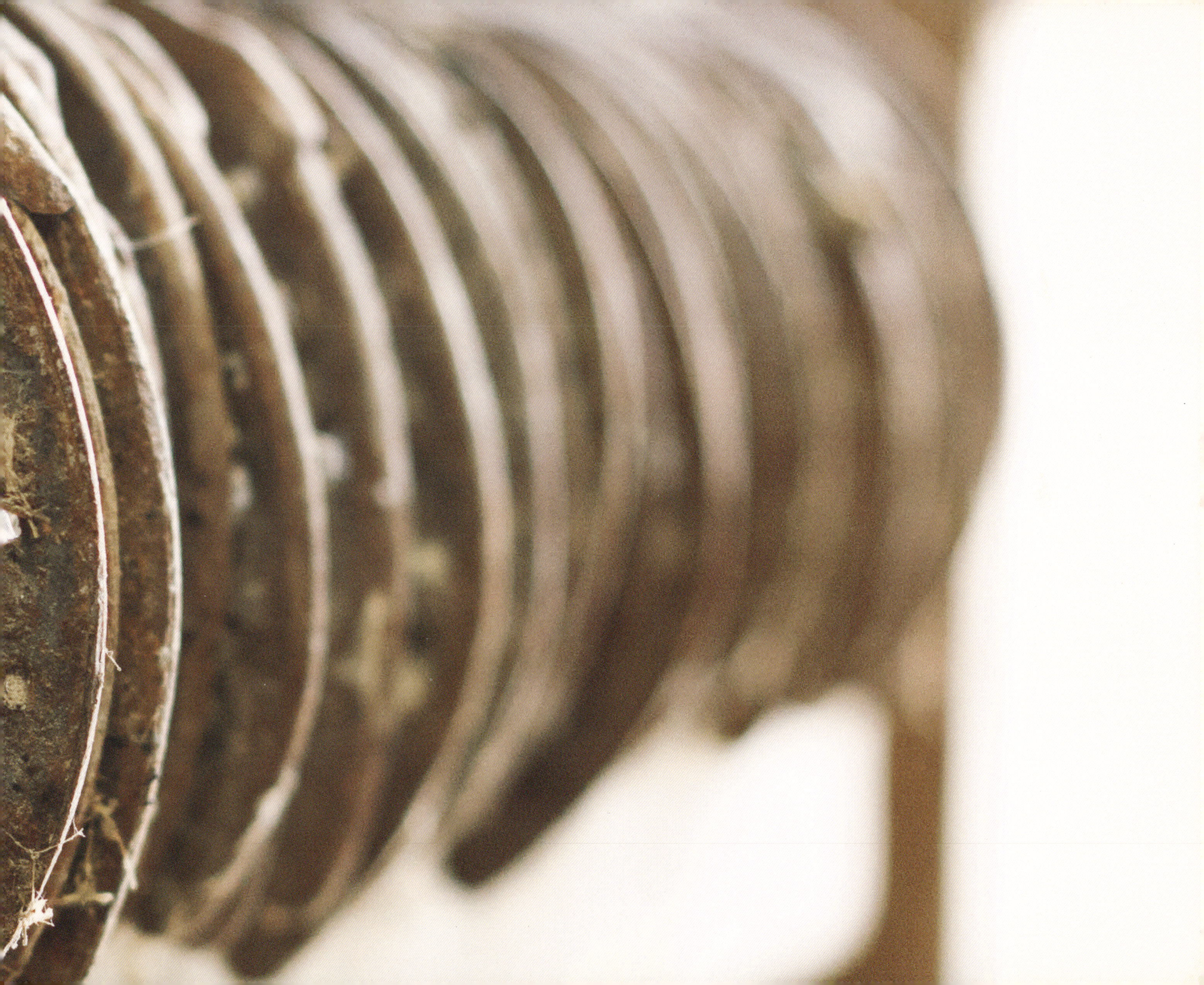

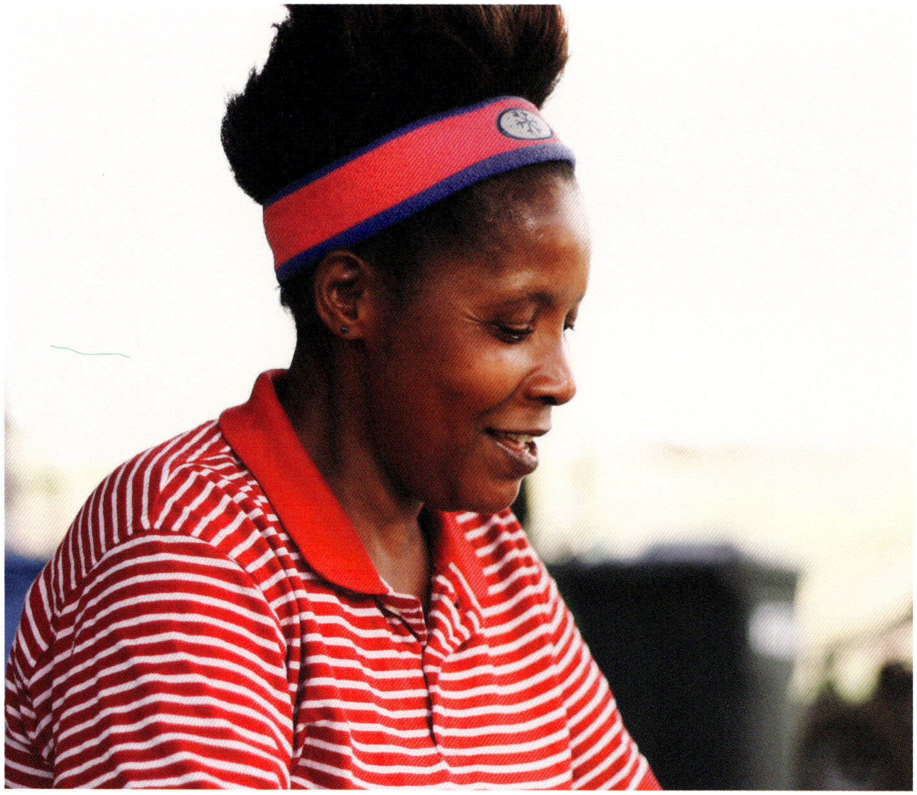

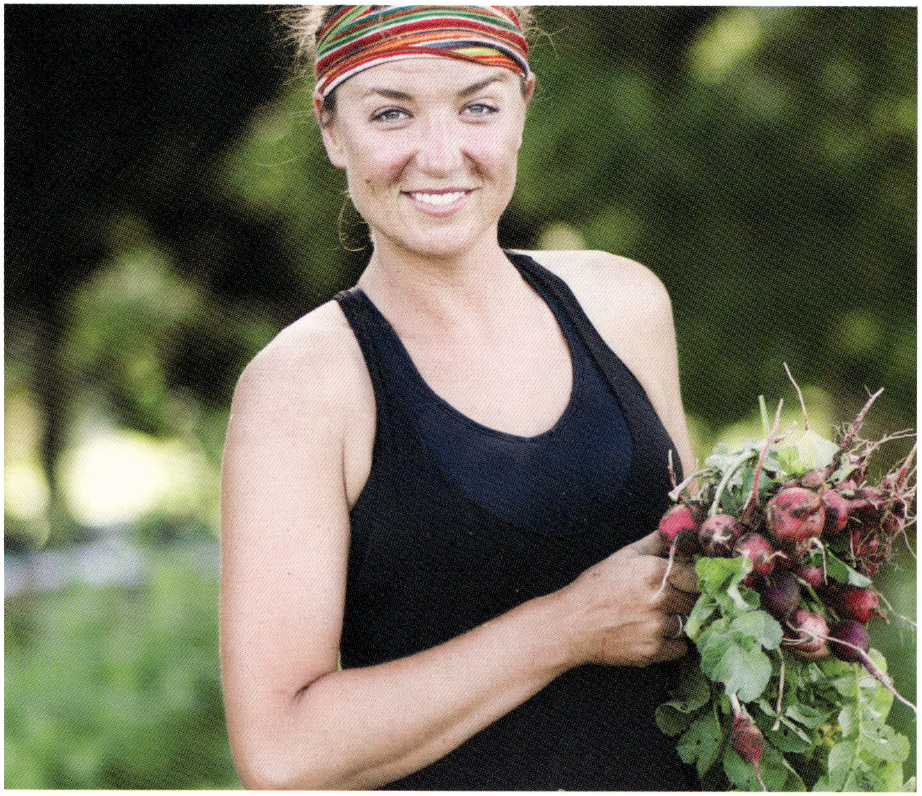

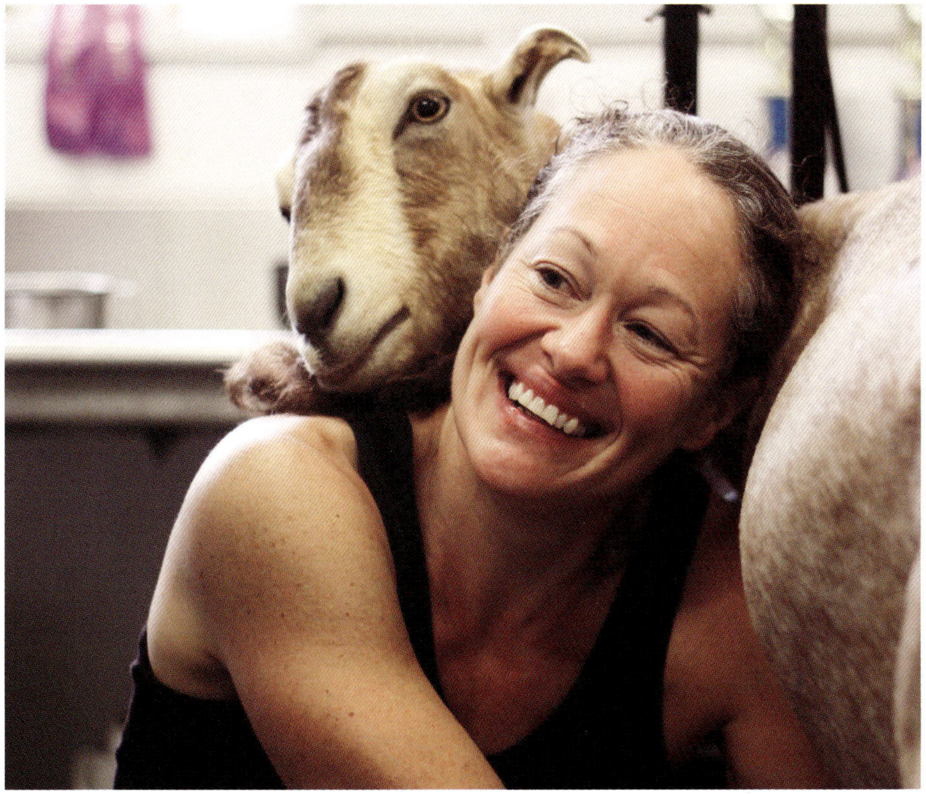

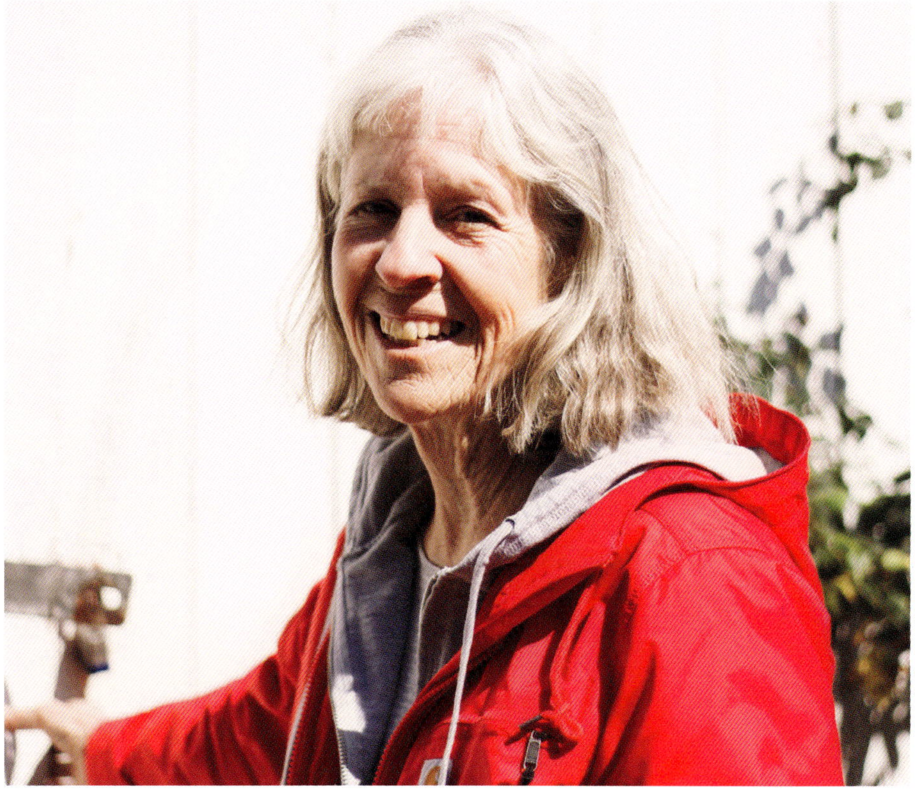

TABLE OF CONTENTS

01. Introduction	**14**	
02. Angelique	**19**	
03. Barbara	**28**	
04. Carol Ann	**43**	
05. Carrie	**55**	
06. Pam	**64**	
07. Reneé	**77**	
08. Sara	**84**	
09. The Documentary	**96**	
10. Marji	**99**	

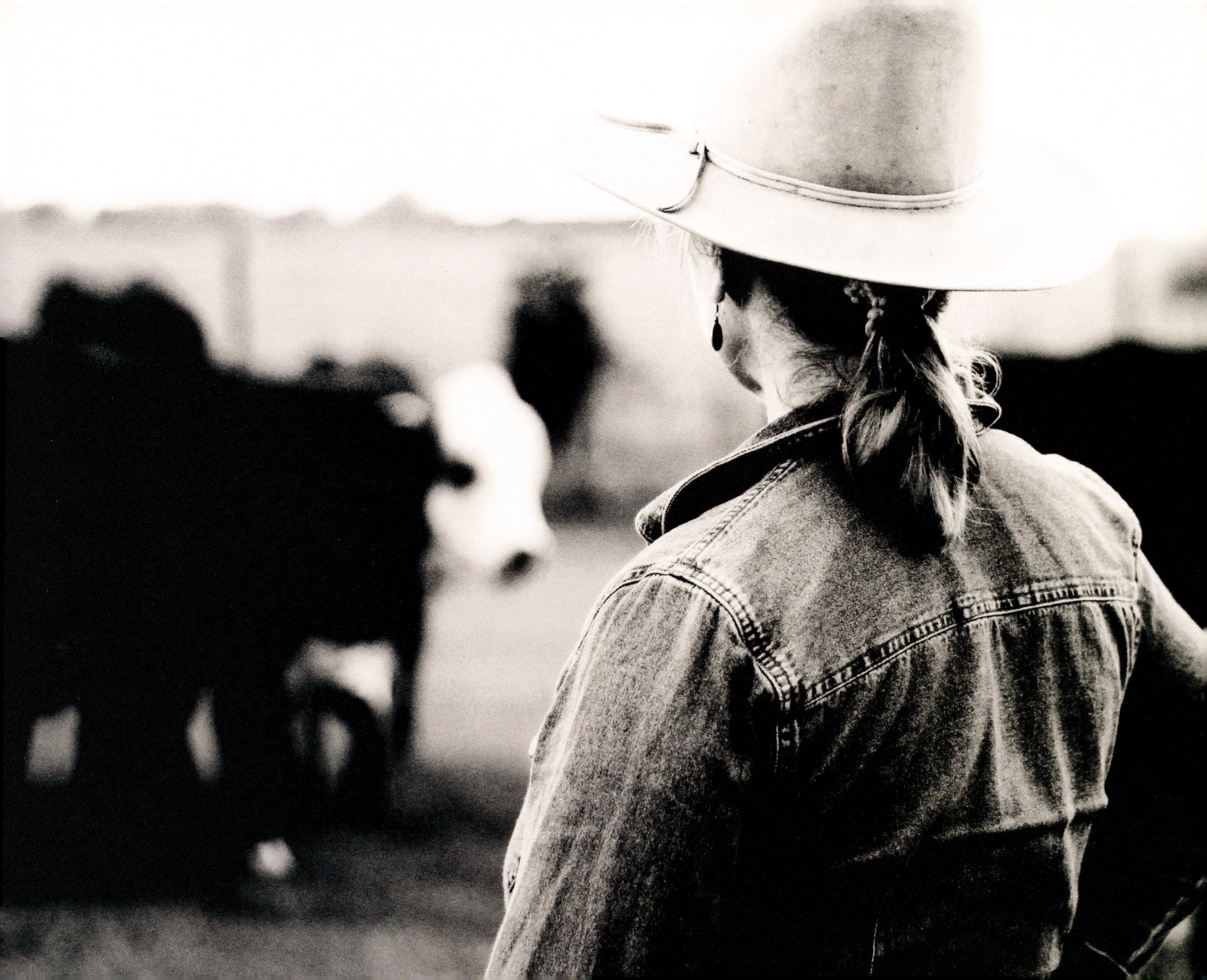

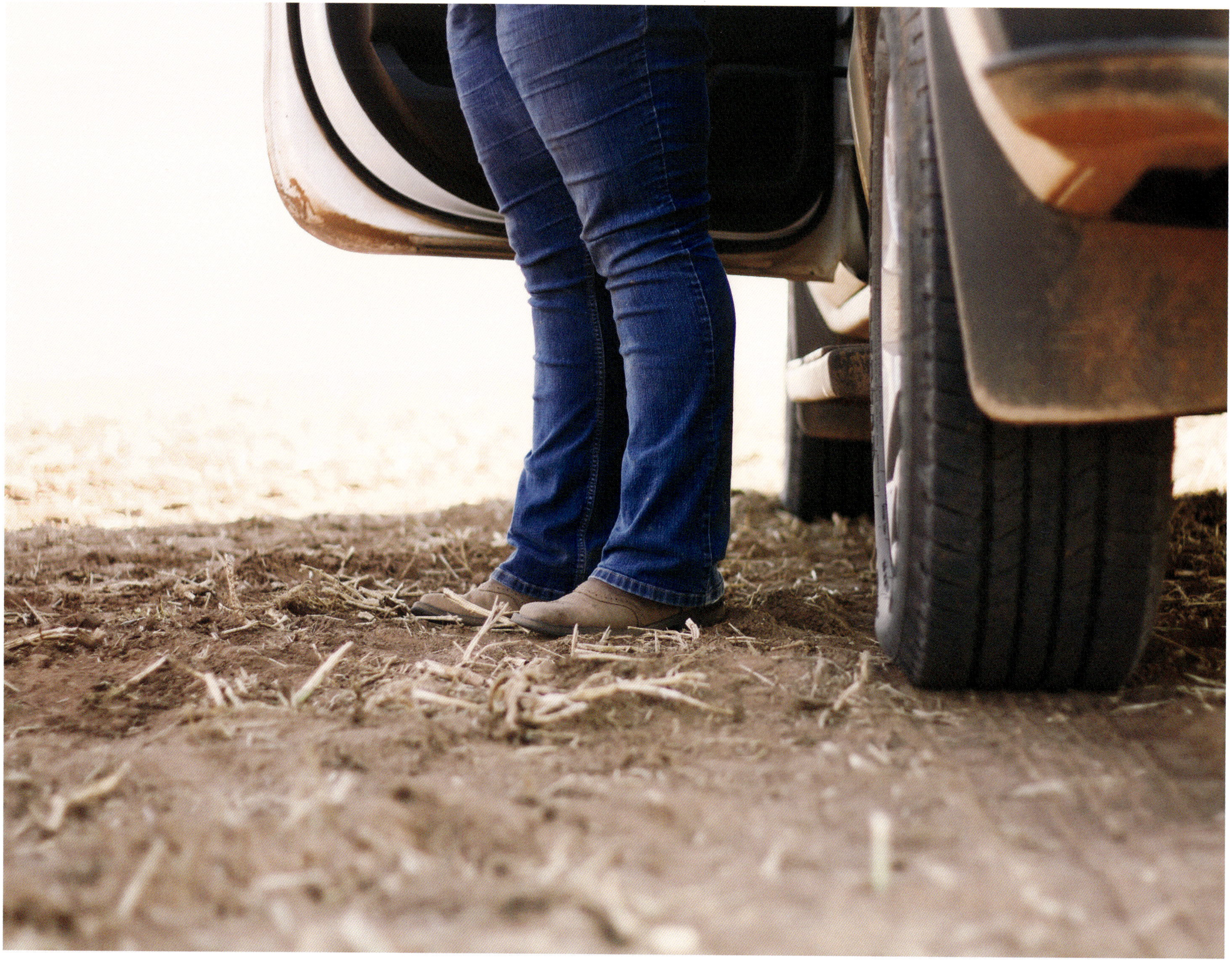

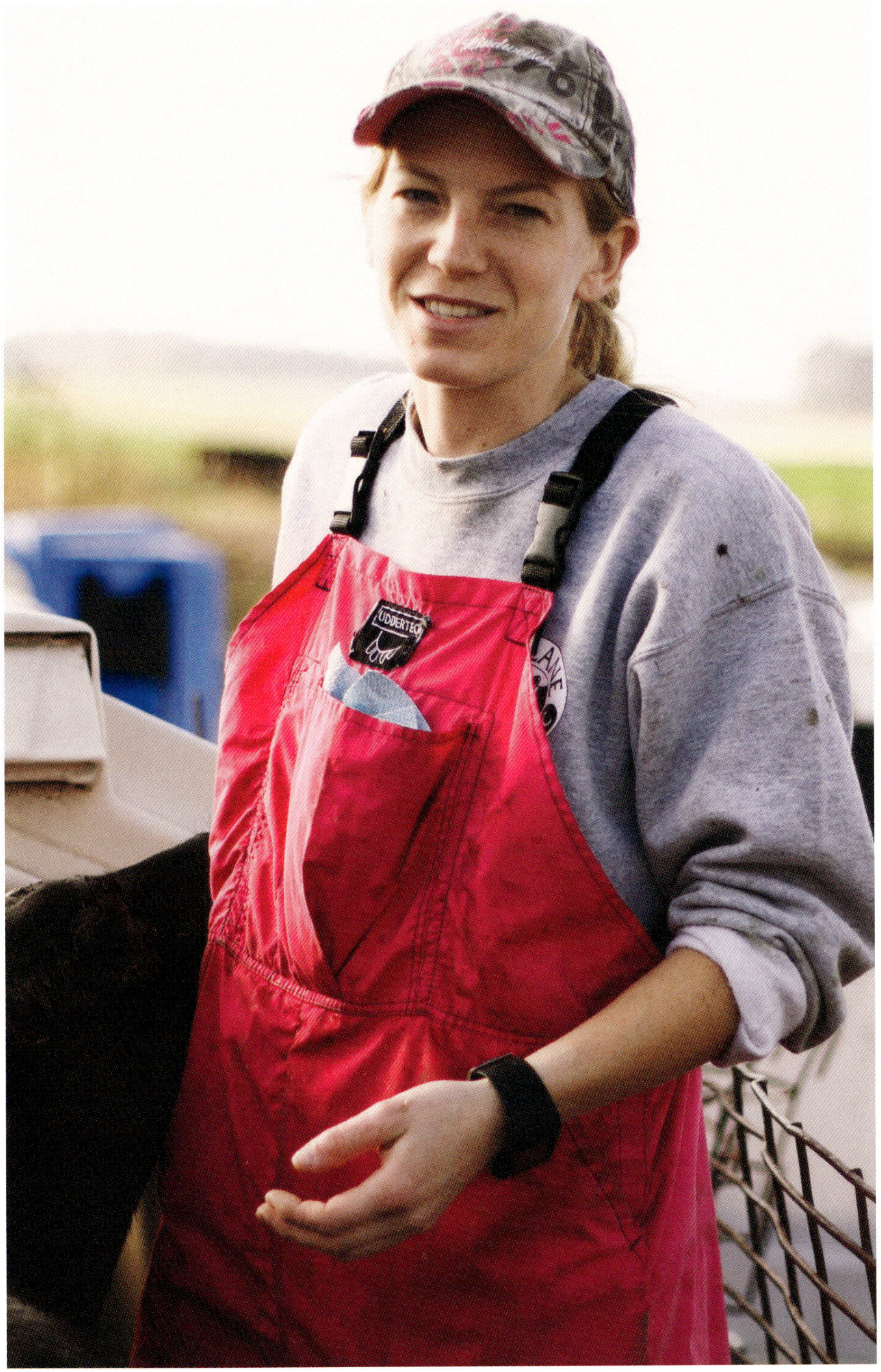

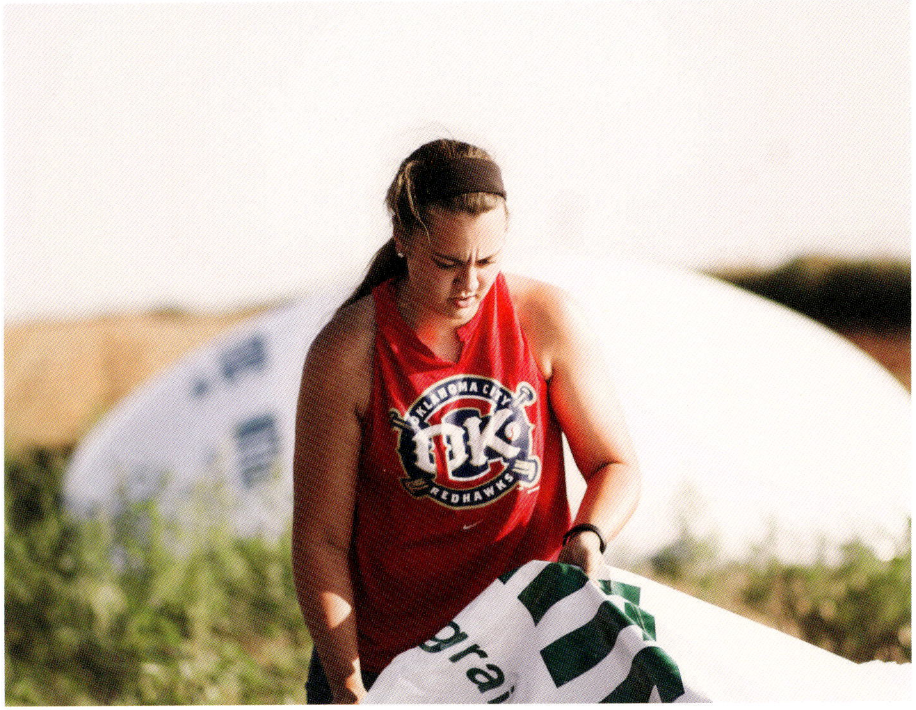
OKLAHOMA CITY
REDHAWKS

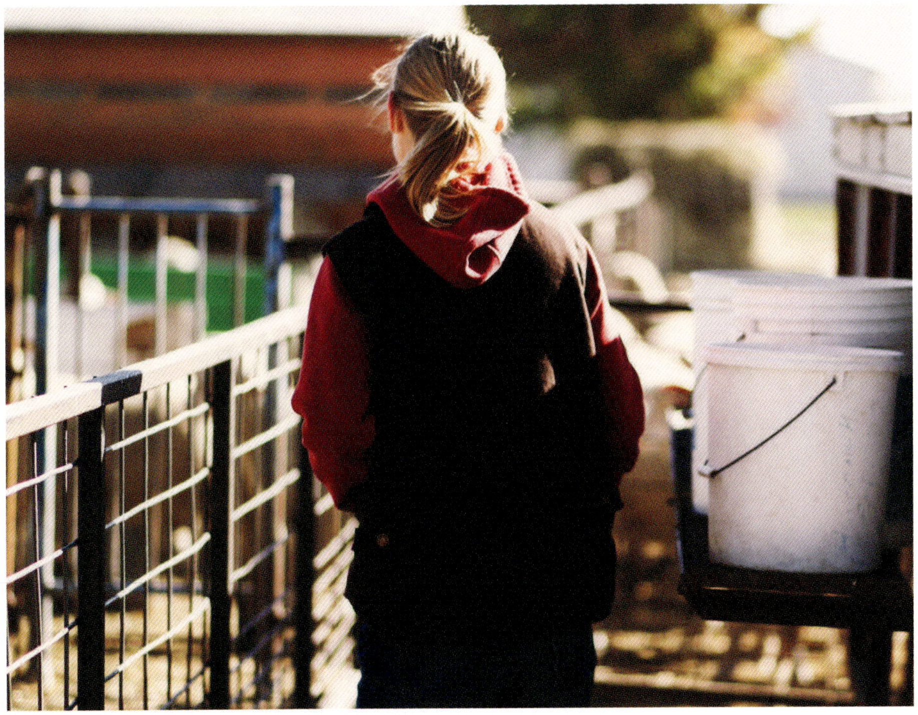

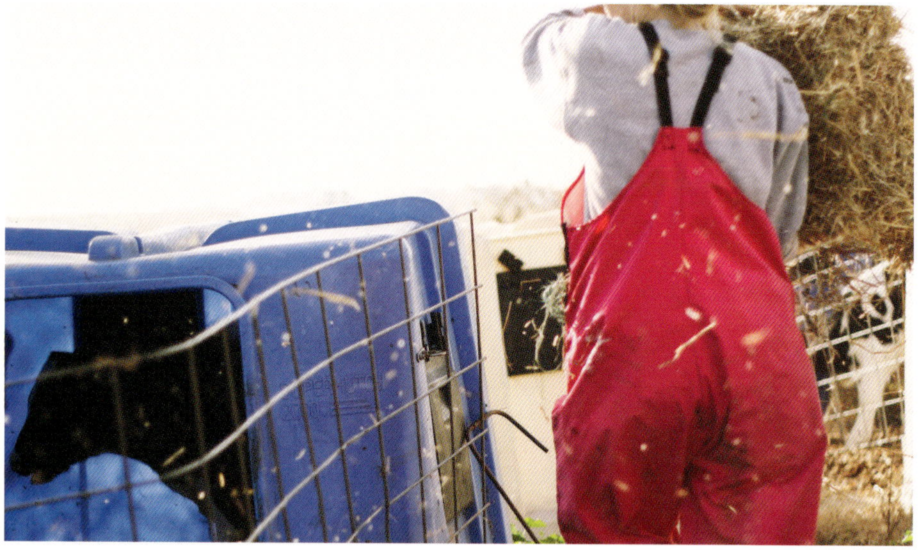

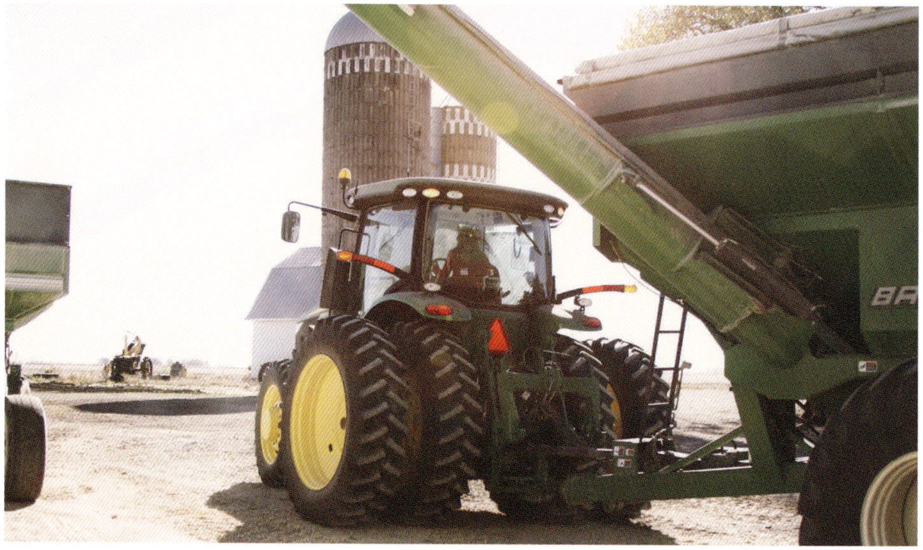

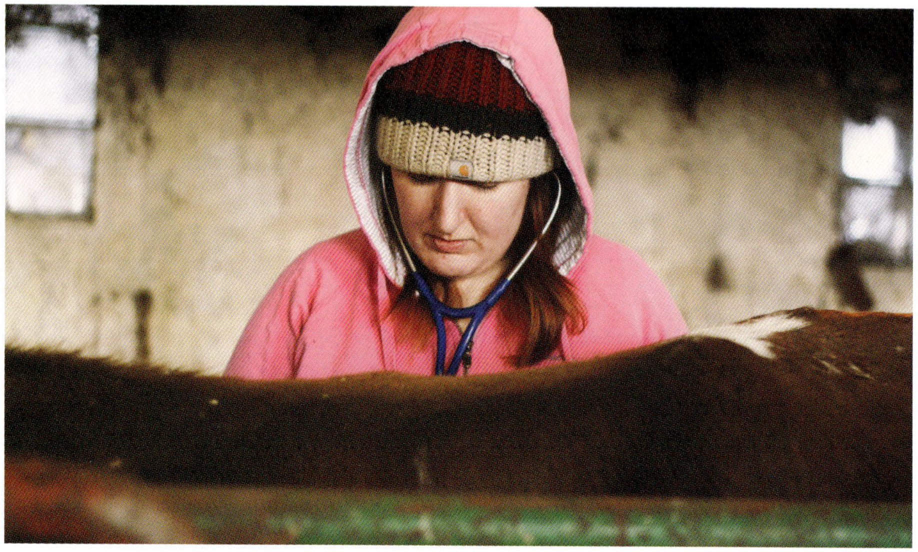

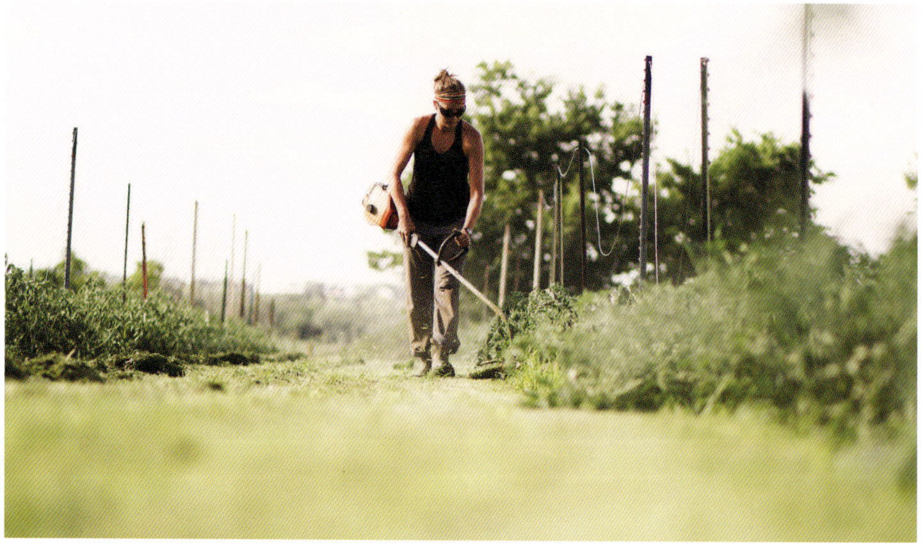

13

FARMHER
SHINING A LIGHT ON WOMEN IN AGRICULTURE

Women have always been an important but mostly unseen aspect of agriculture. In recent years, women have risen to the forefront of agriculture in so many ways; as owner/operators, landowners, workers, mentors and much more.

FarmHer was founded in 2013 to begin to change the image of agriculture—to include women in that image through photographs and stories. It quickly became clear that women in agriculture not only appreciated FarmHer, but they needed it. And they asked for more. Today, FarmHer has grown into not just a gallery of images that are changing the way people perceive a farmer or rancher, but also into an online community built just for women in agriculture. It is a place where this imperative group of women can experience stories about others like them, a place where they can connect with and learn from others in a safe and positive environment and a place where women rise to the forefront of agriculture.

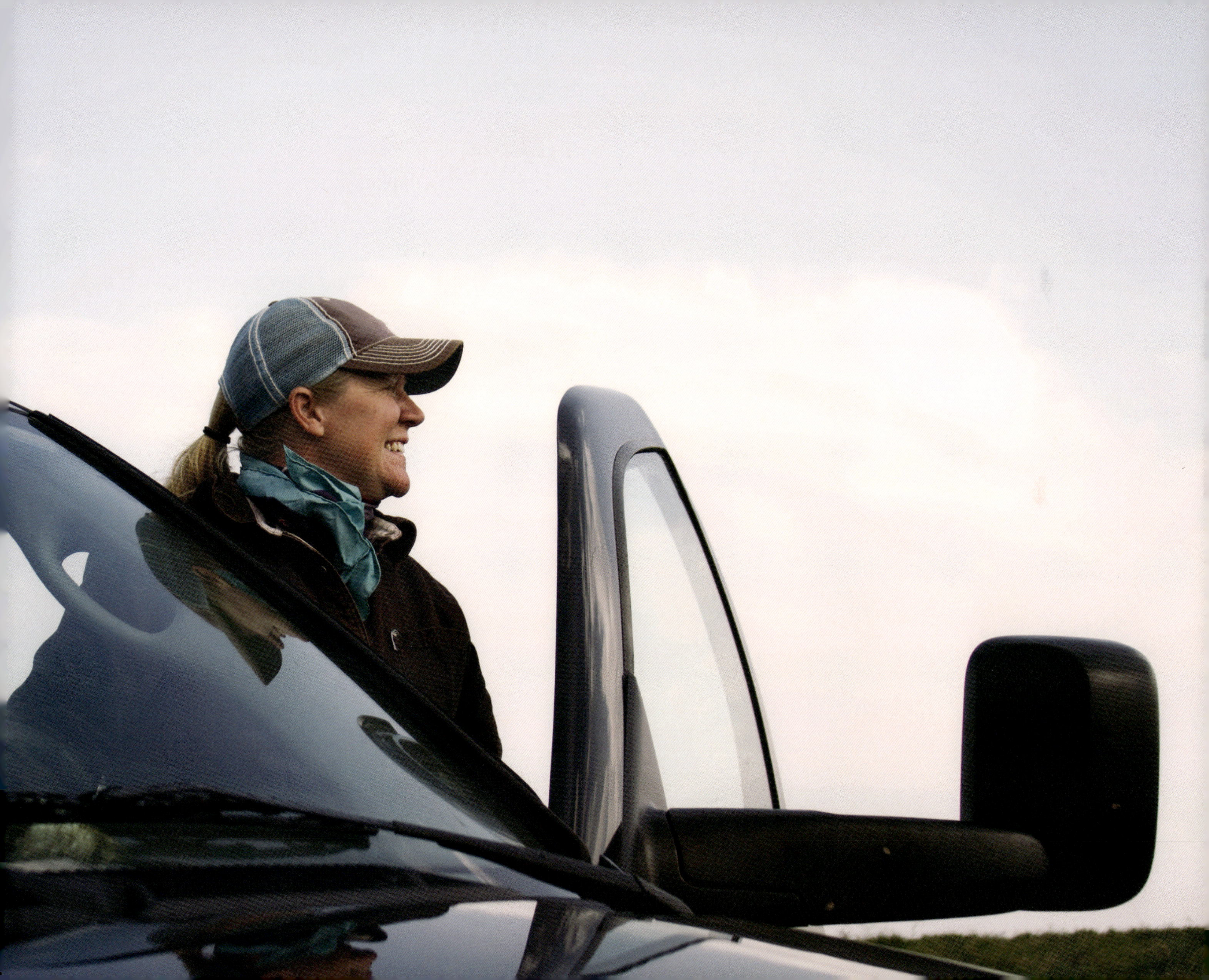

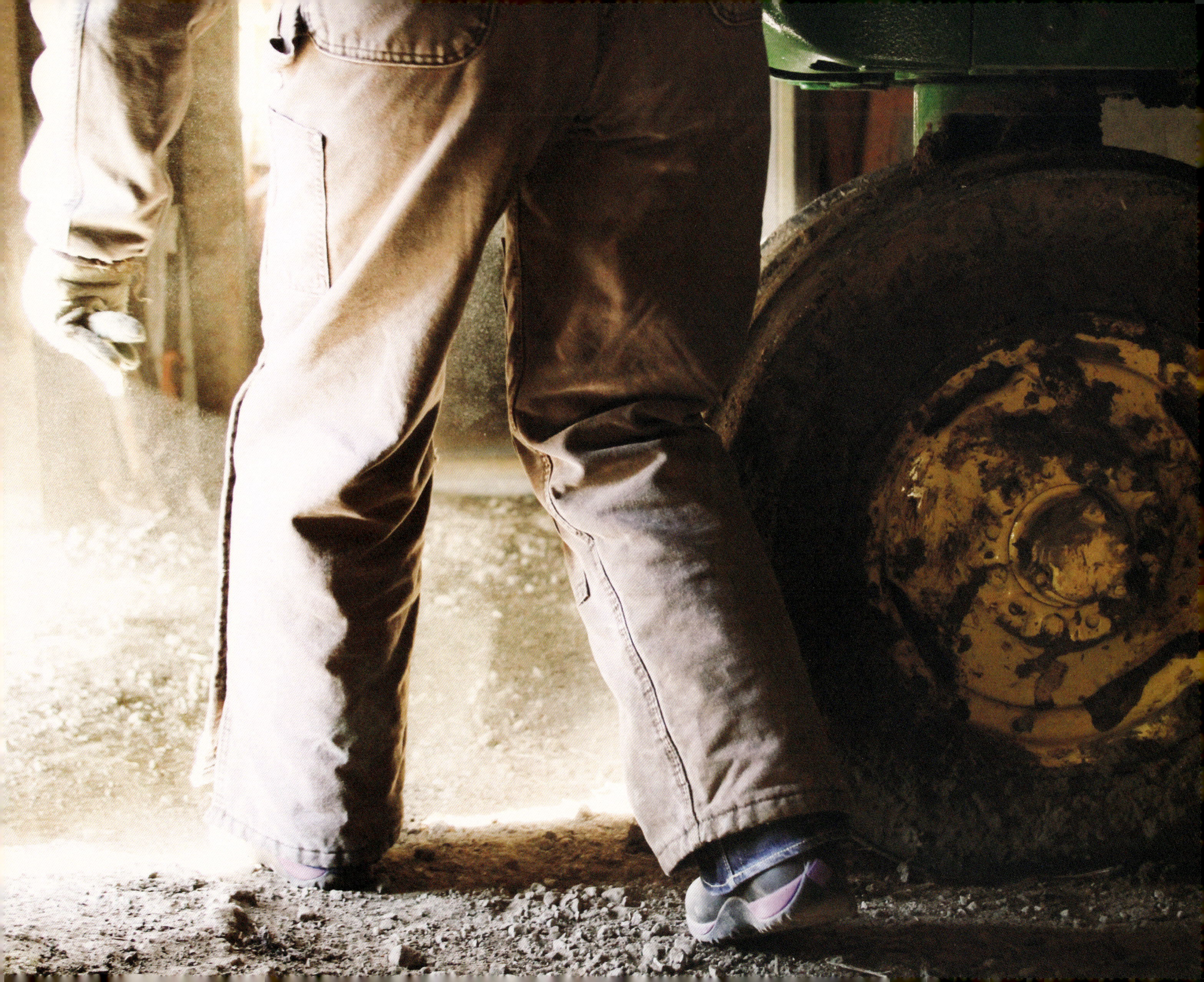

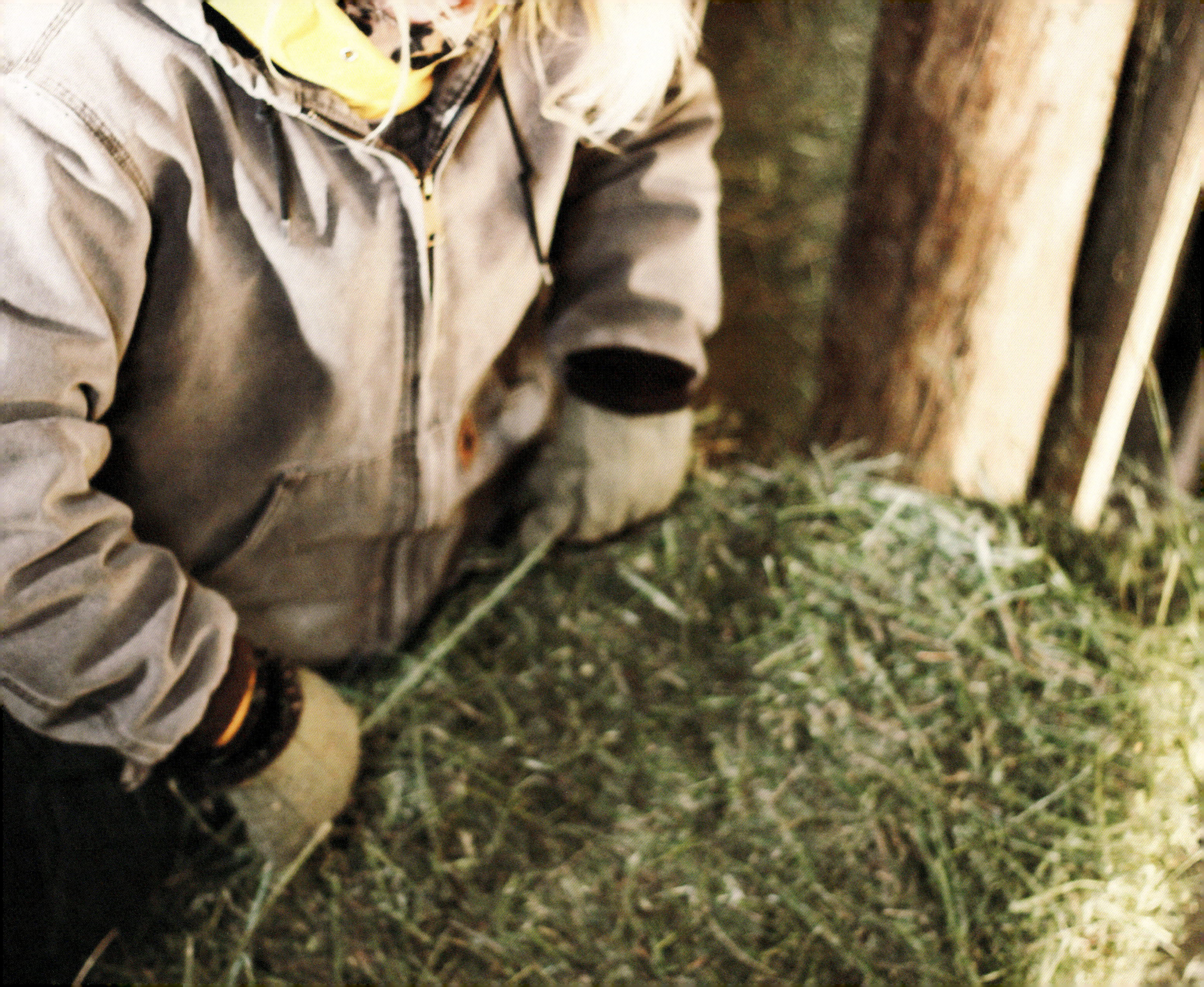

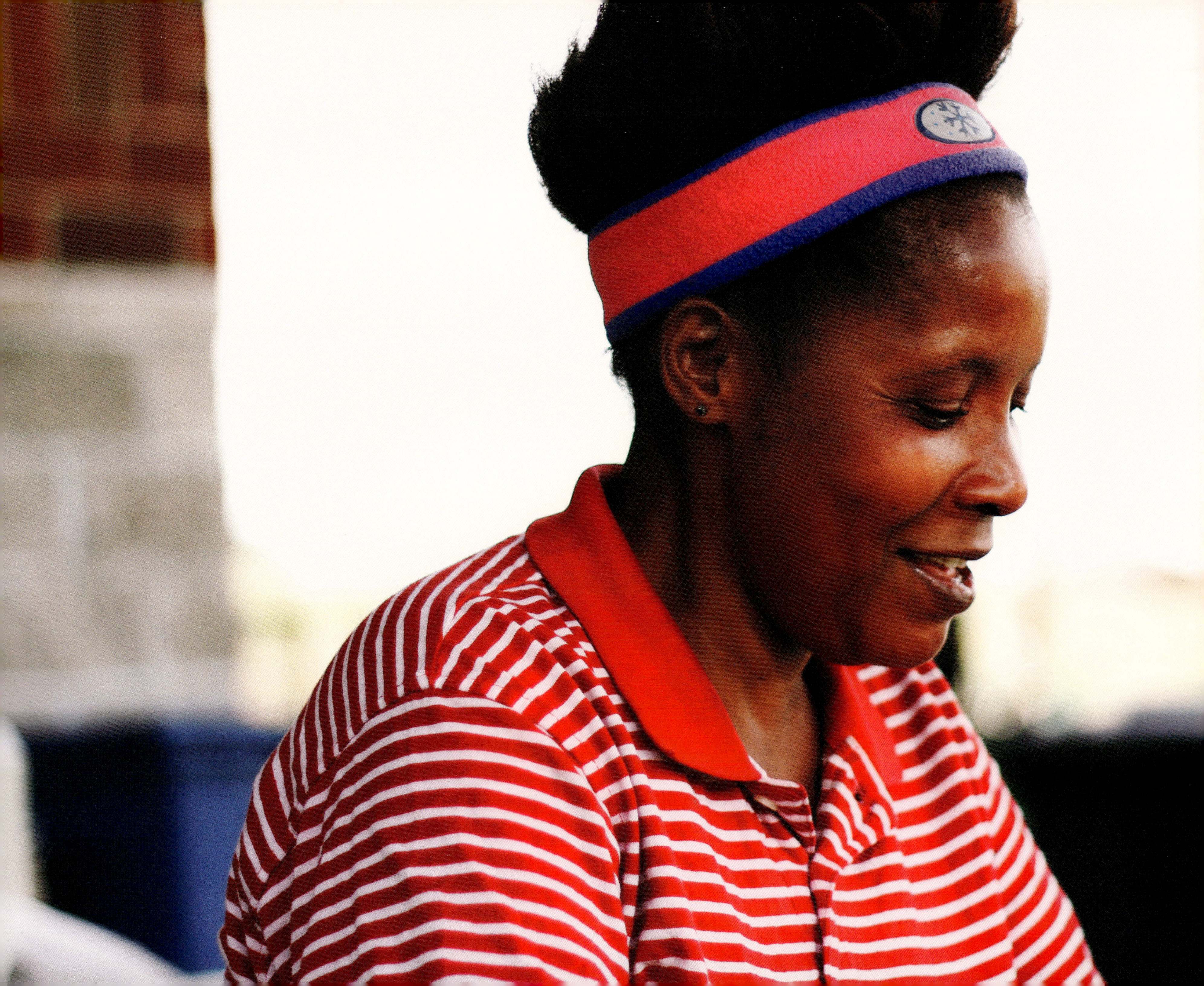

ANGELIQUE
GLOBAL GREENS FARM • IOWA

Angelique Hakuzimana fled war in her home country of Rwanda, traveling to the Congo, Central African Republic, and then making her way to the United States in 2009. She has faced adversity and lost many family members to the war in her native land, but she never gave up on her love of farming. Her father taught her how to farm when she was a child and she wants to carry on the tradition.

She is able to farm by way of the Lutheran Services in Iowa refugee program and has her own plot of land that she tends to at the Global Greens Farm in West Des Moines. She works other jobs to make a living, but her passion is farming.

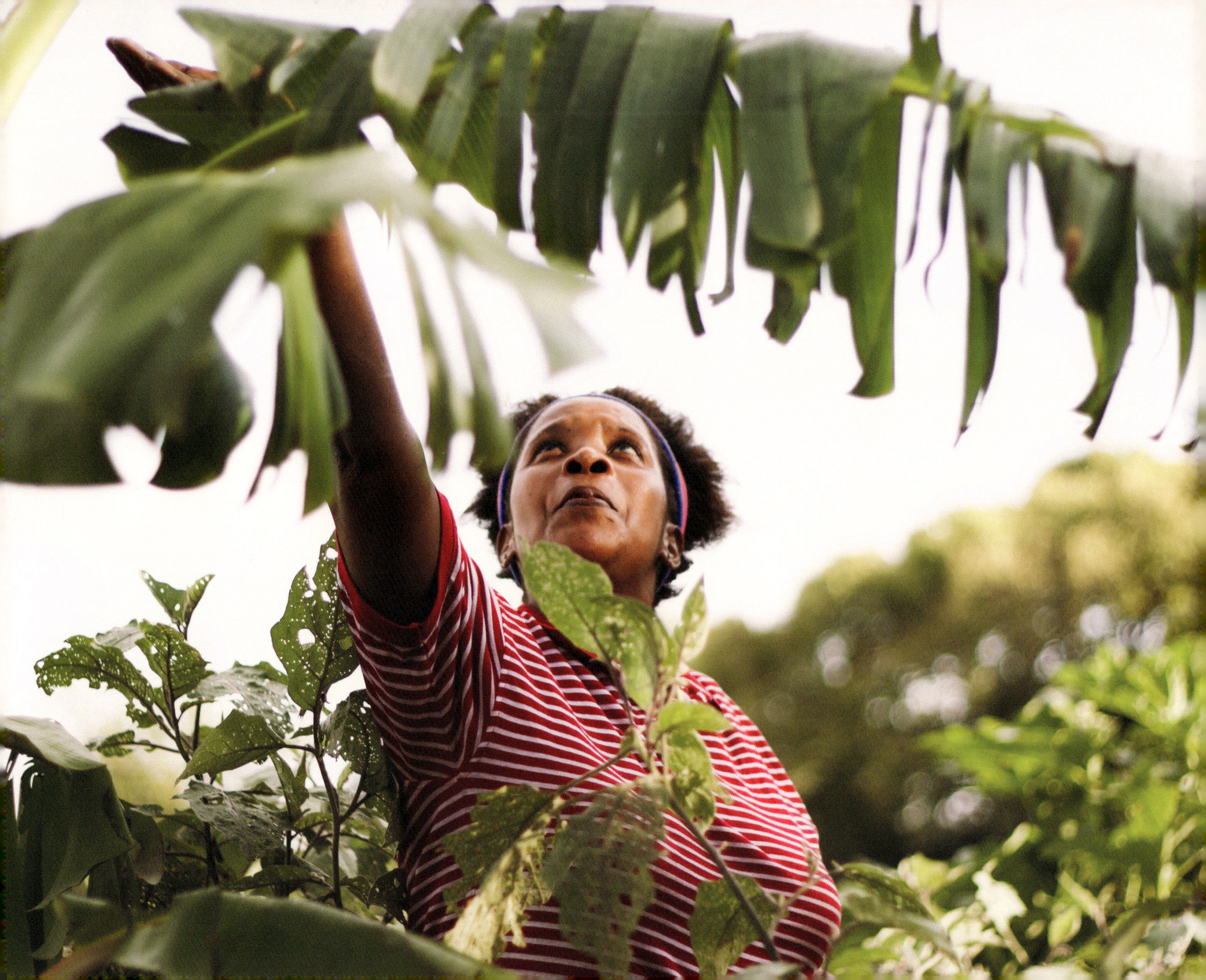

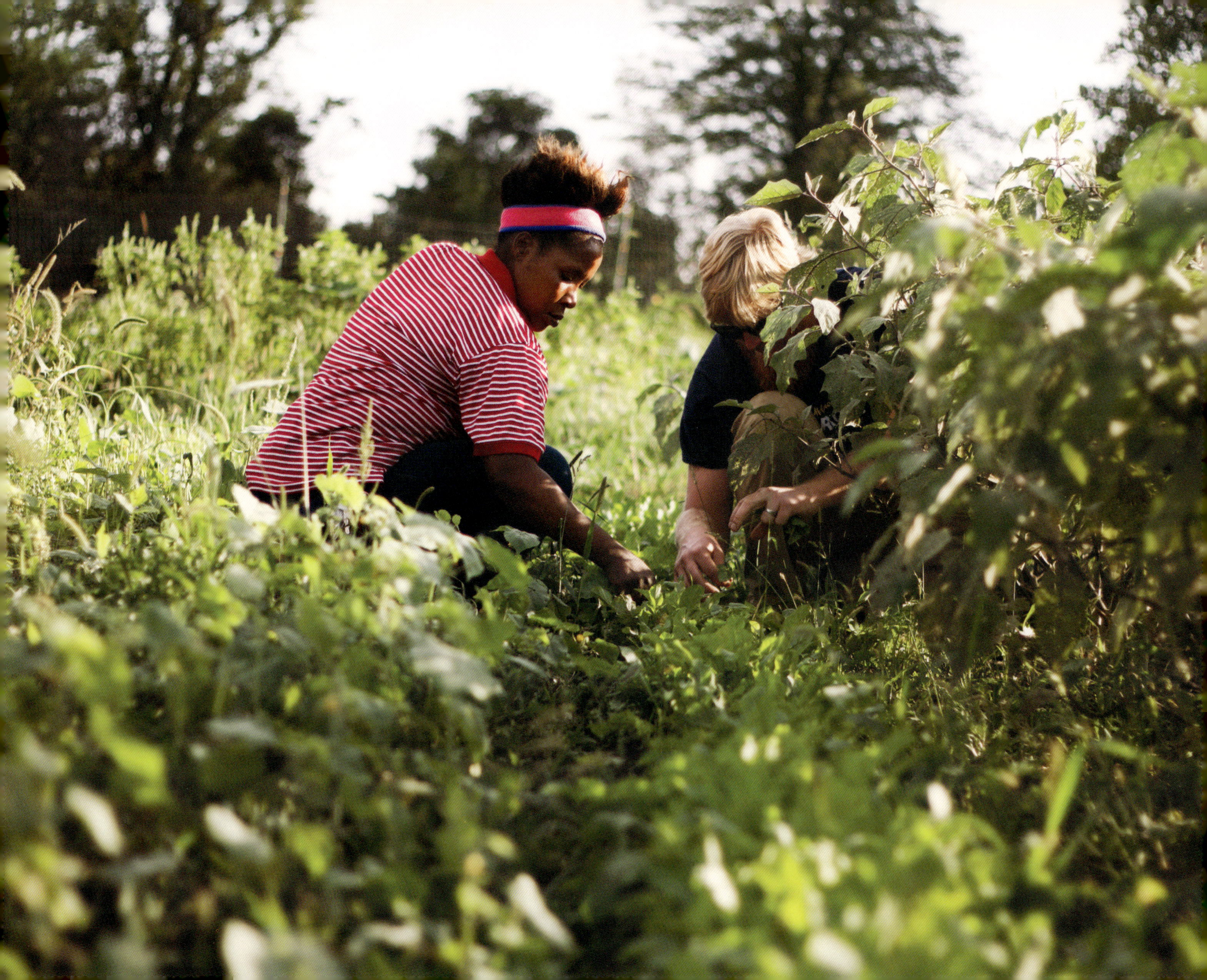

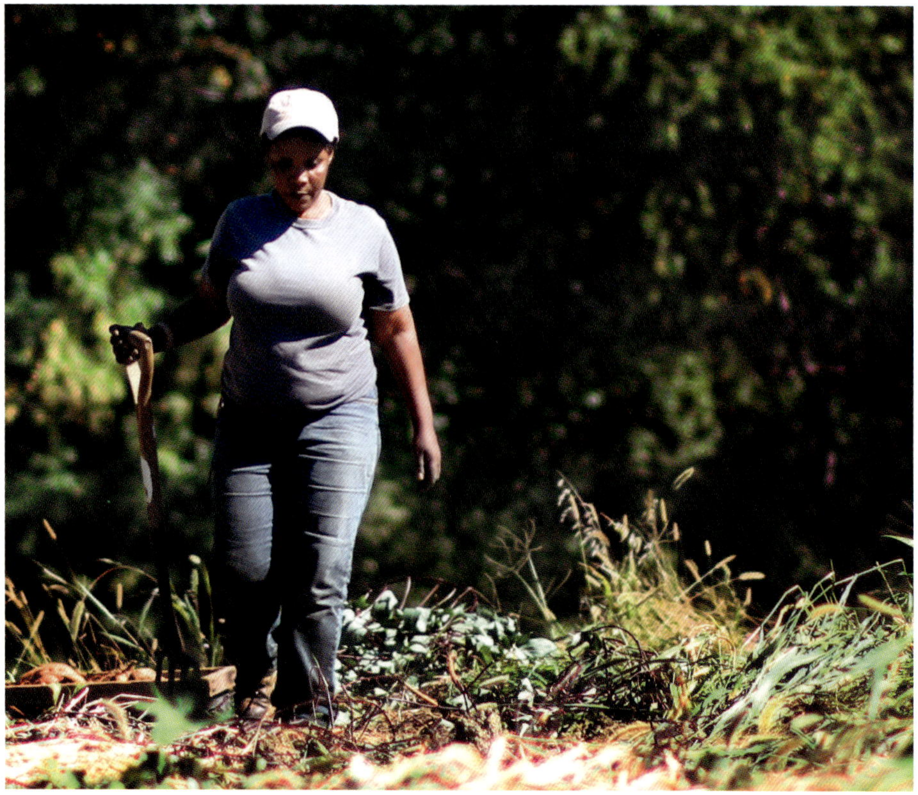

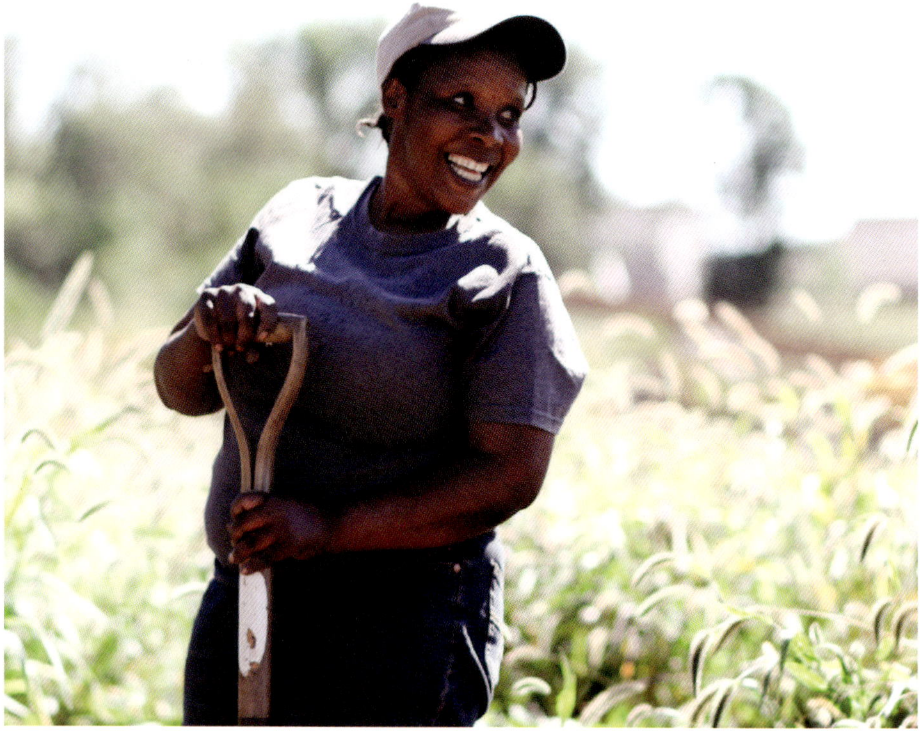

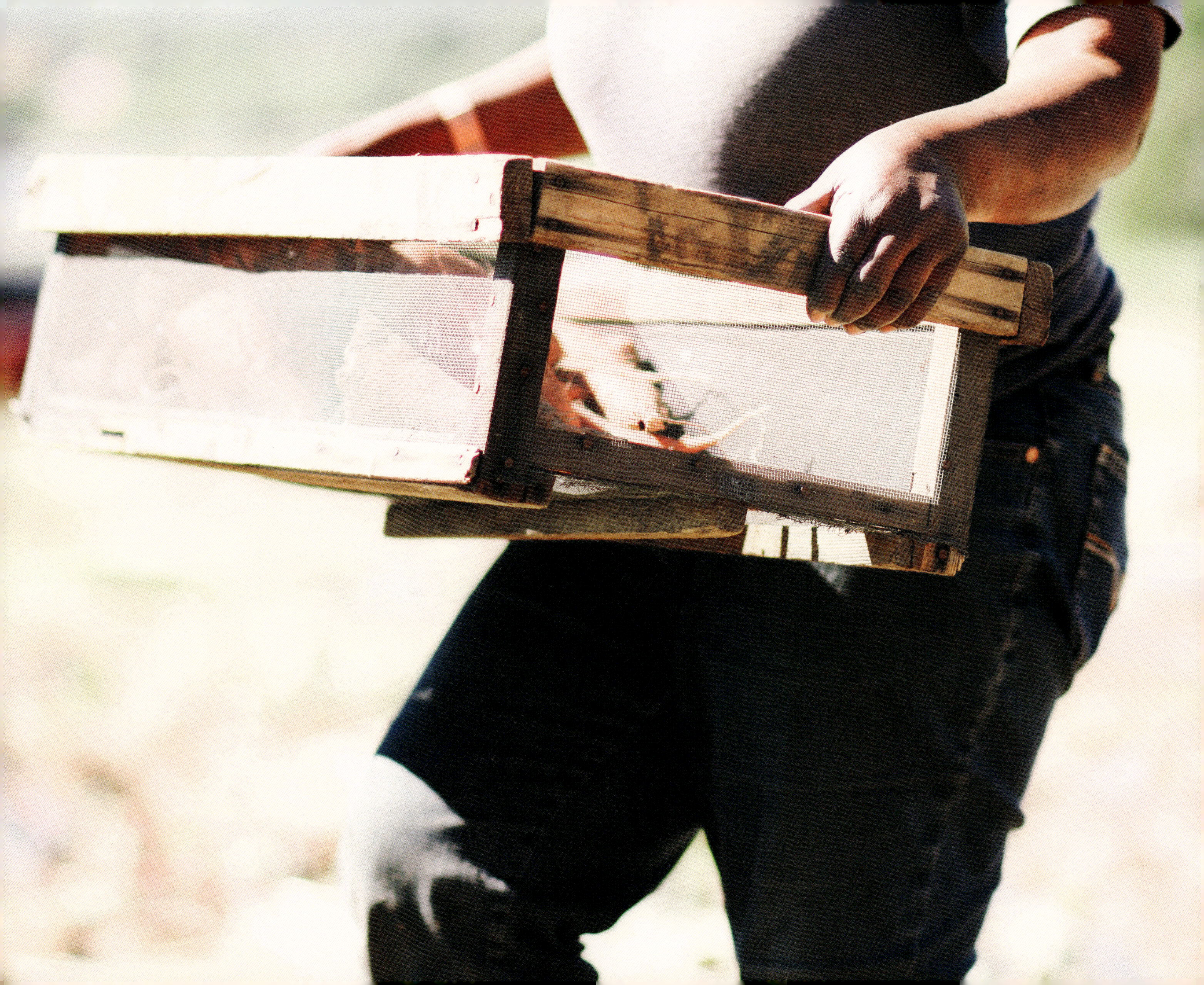

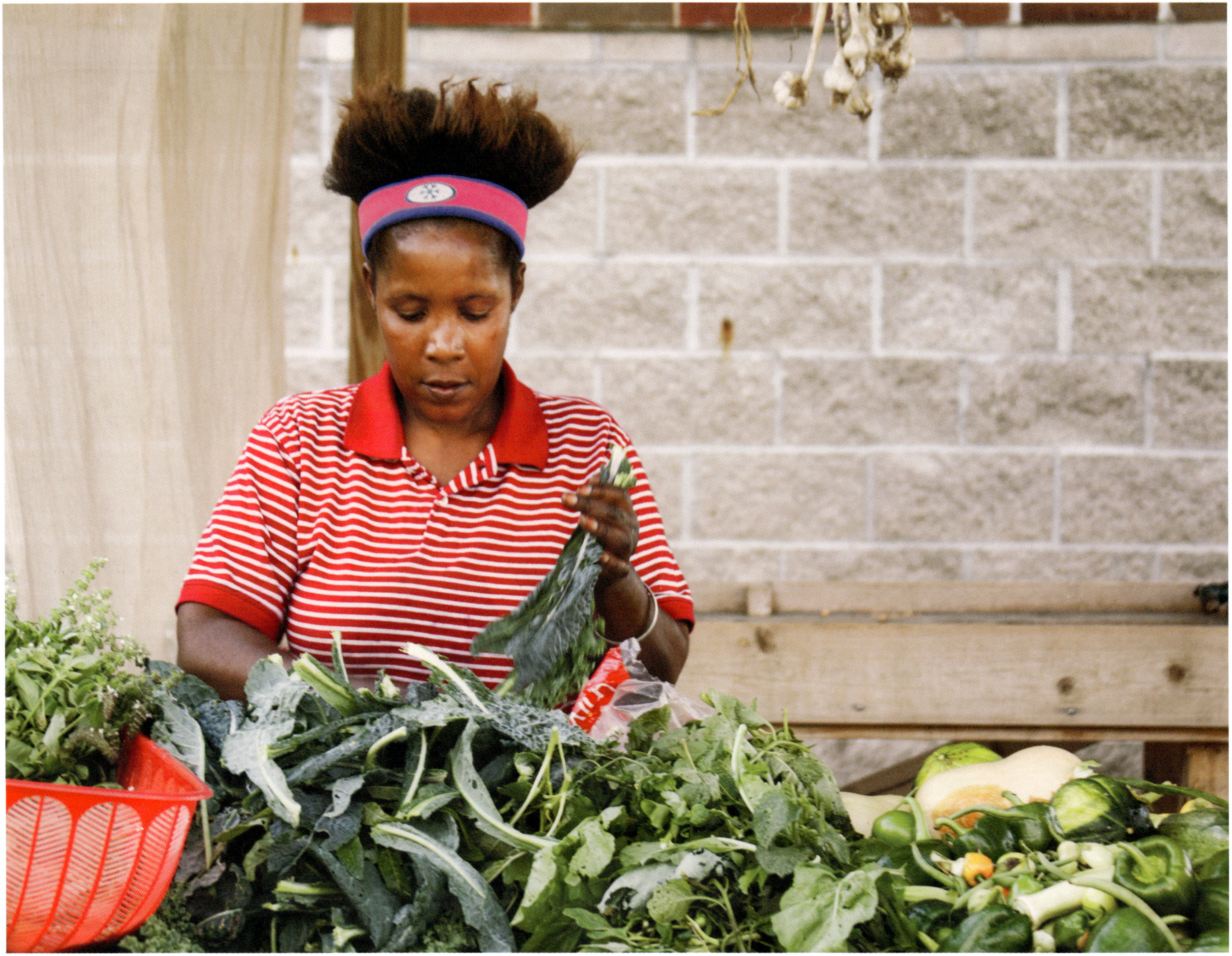

25

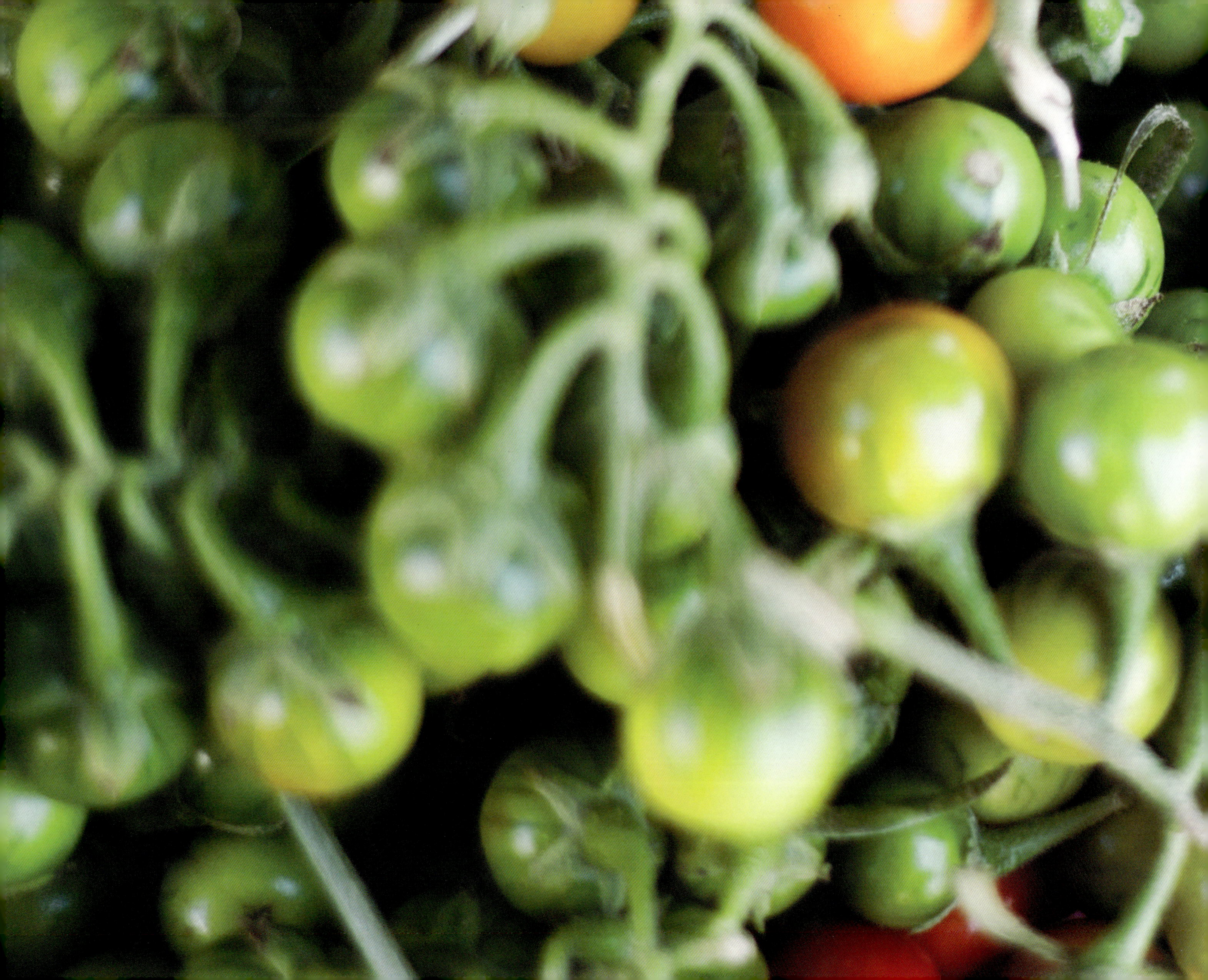

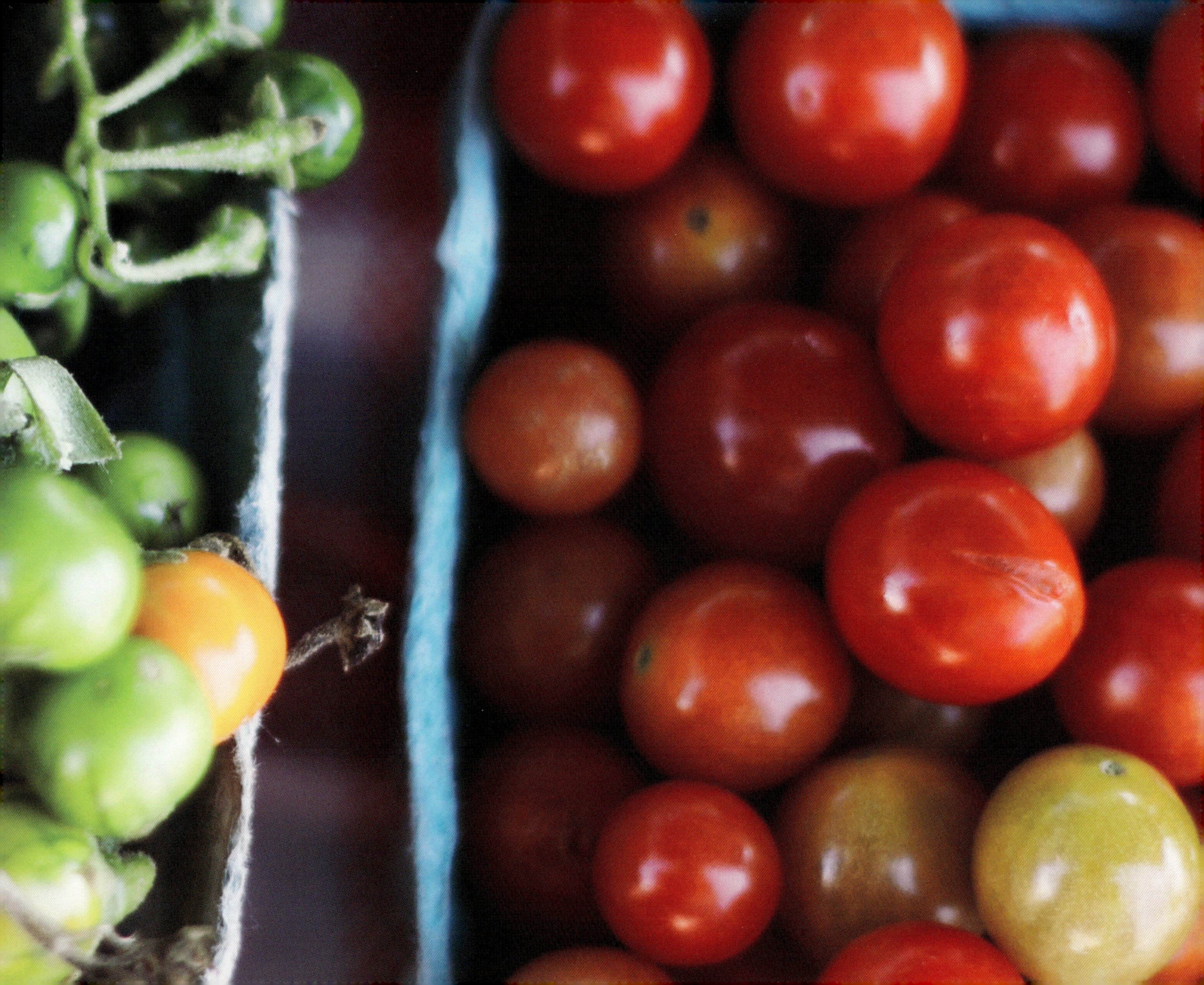

BARBARA

MAZUREK FAMILY RANCH • TEXAS

Barbara Mazurek was born and raised a rancher, and says it's in her blood. Located in southwest Texas in the Hill Country, the Mazurek Family Ranch has been in the family since the mid-1800s.

The ranch includes a base Angora goat herd and several hundred boer goats. Her animals are shown competitively by her grandchildren as well. At 79, Barbara runs the ranch on her own. Sadly, her husband died in a tragic farming accident more than 20 years ago.

Barbara is also active in many agriculture programs for youth in her community. She's a leader in the industry and an inspiration to other women.

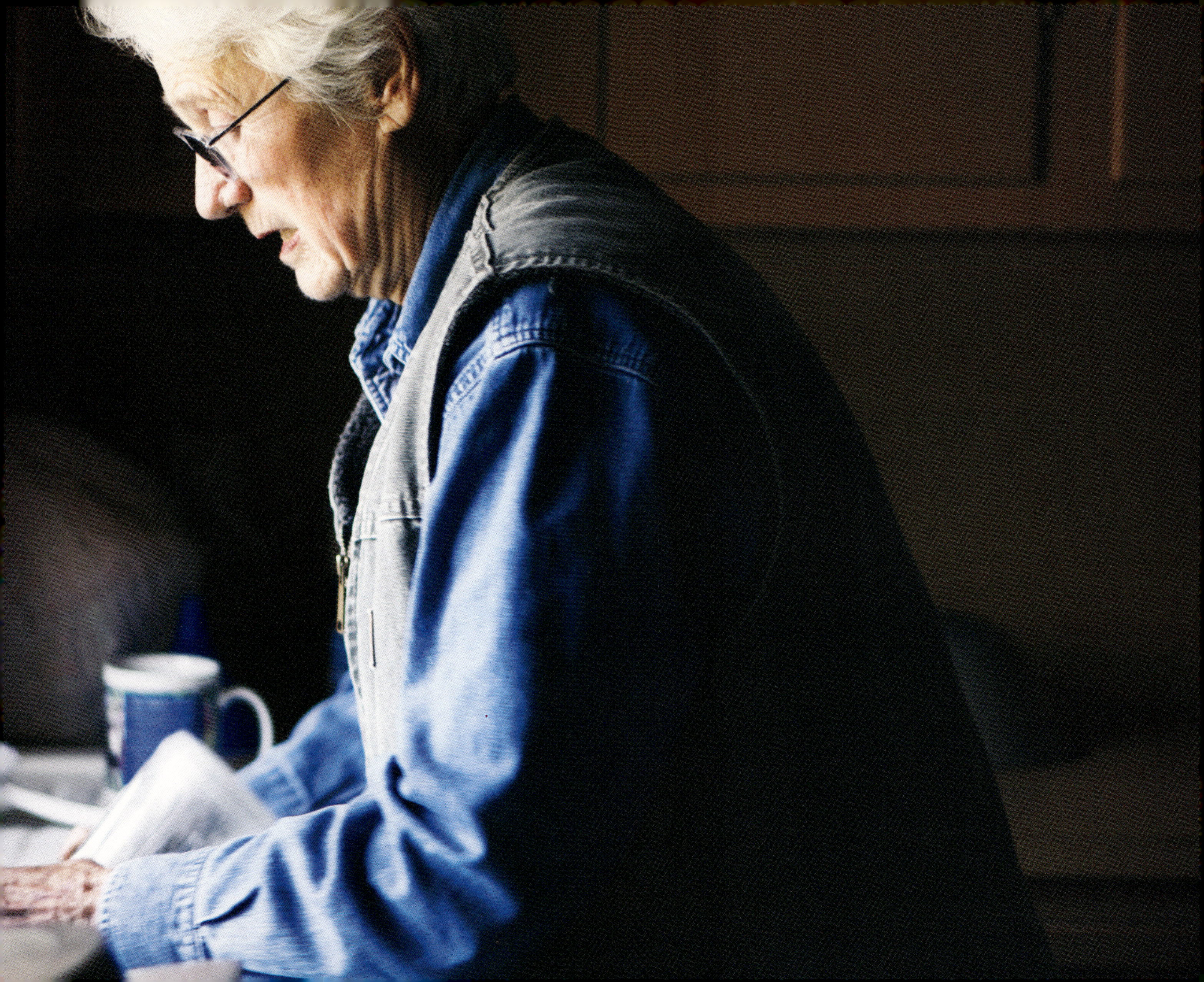

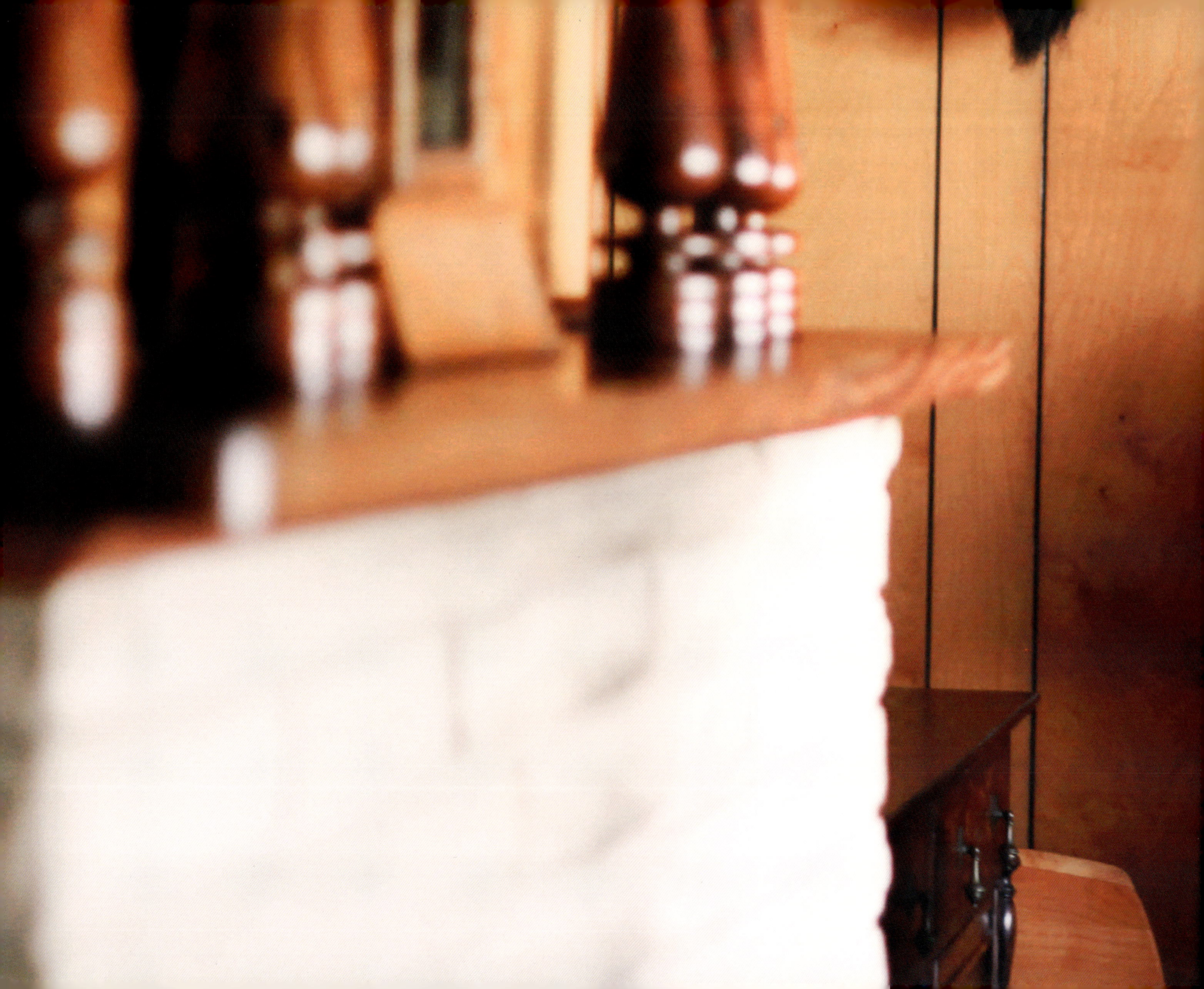

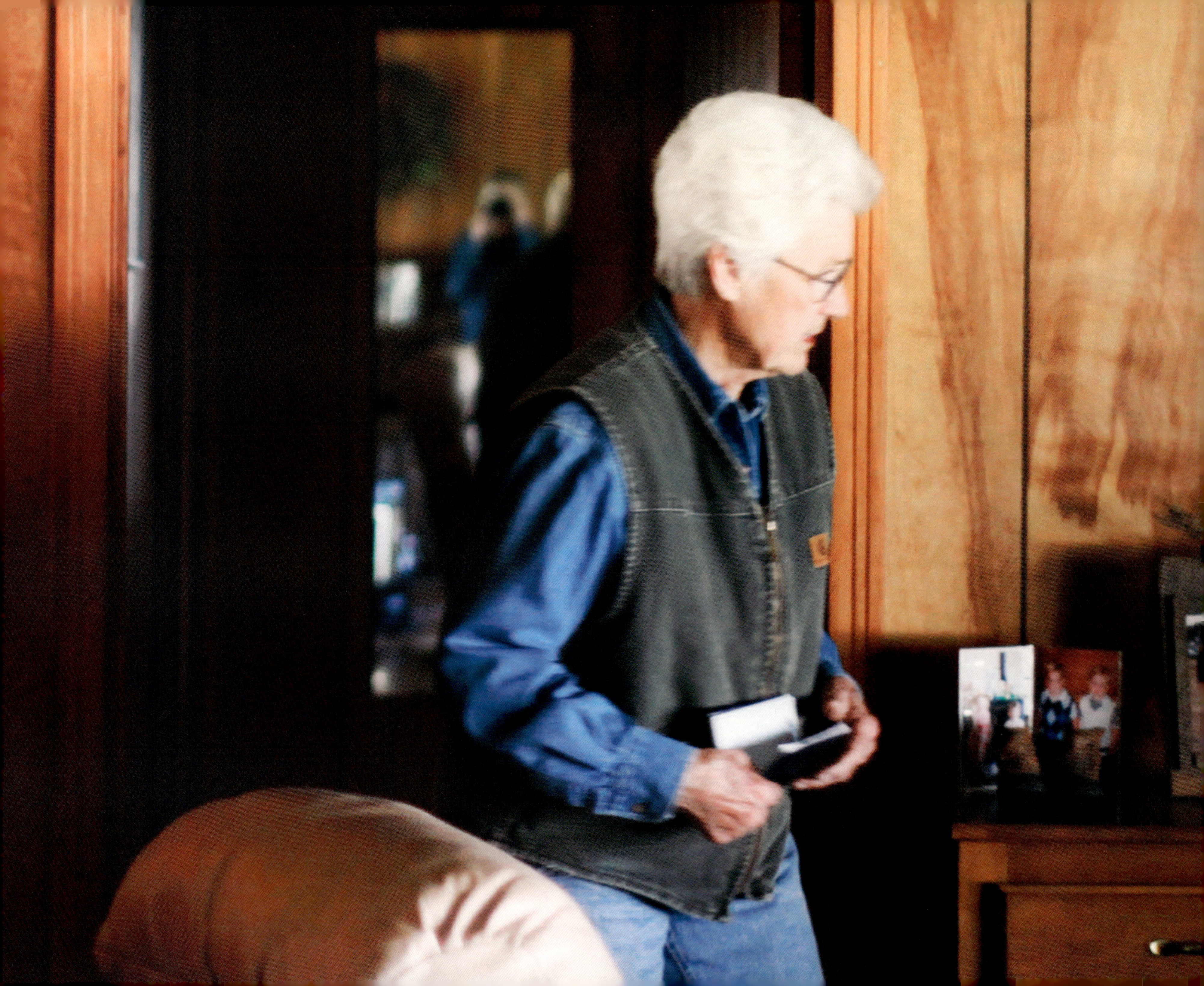

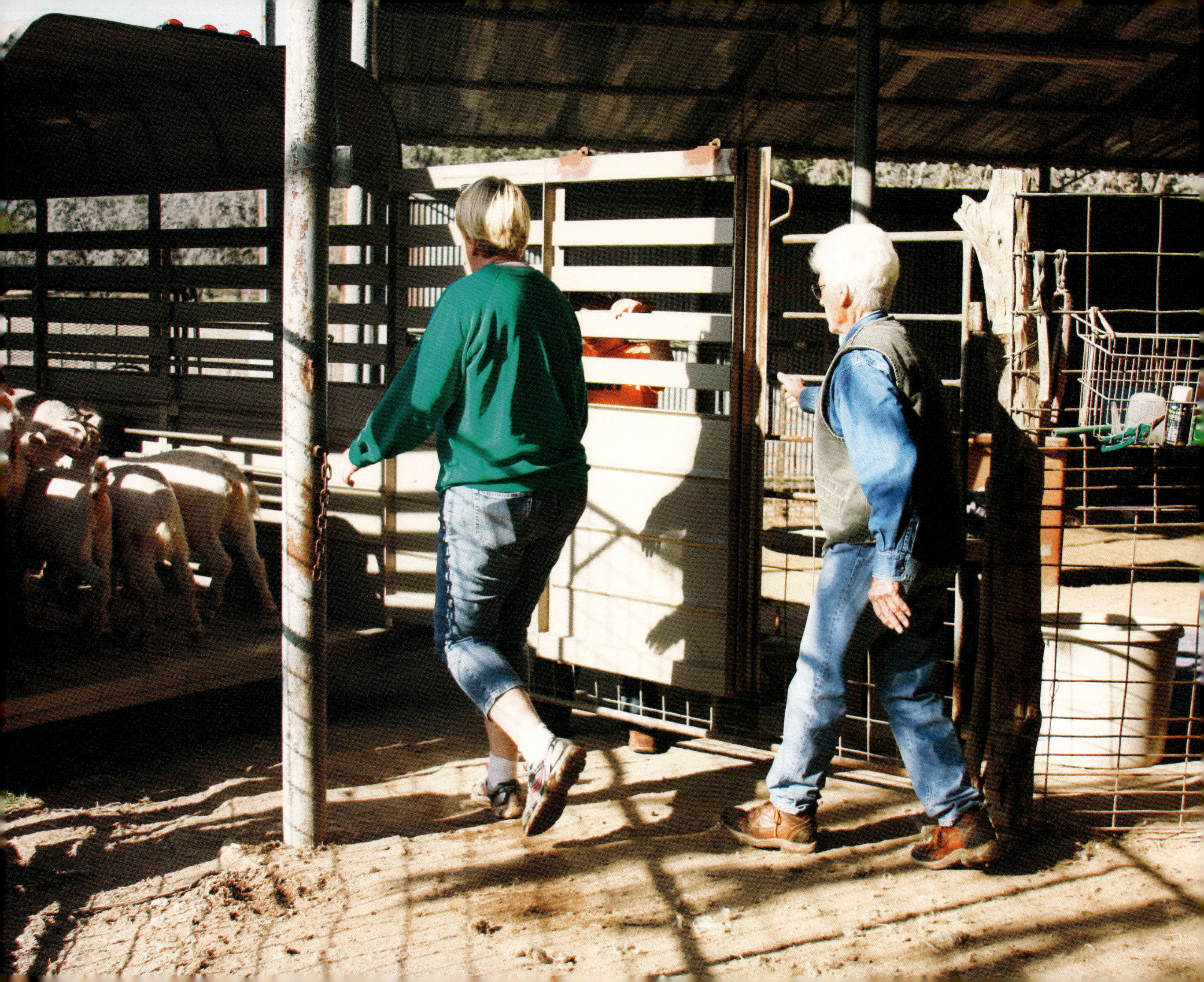

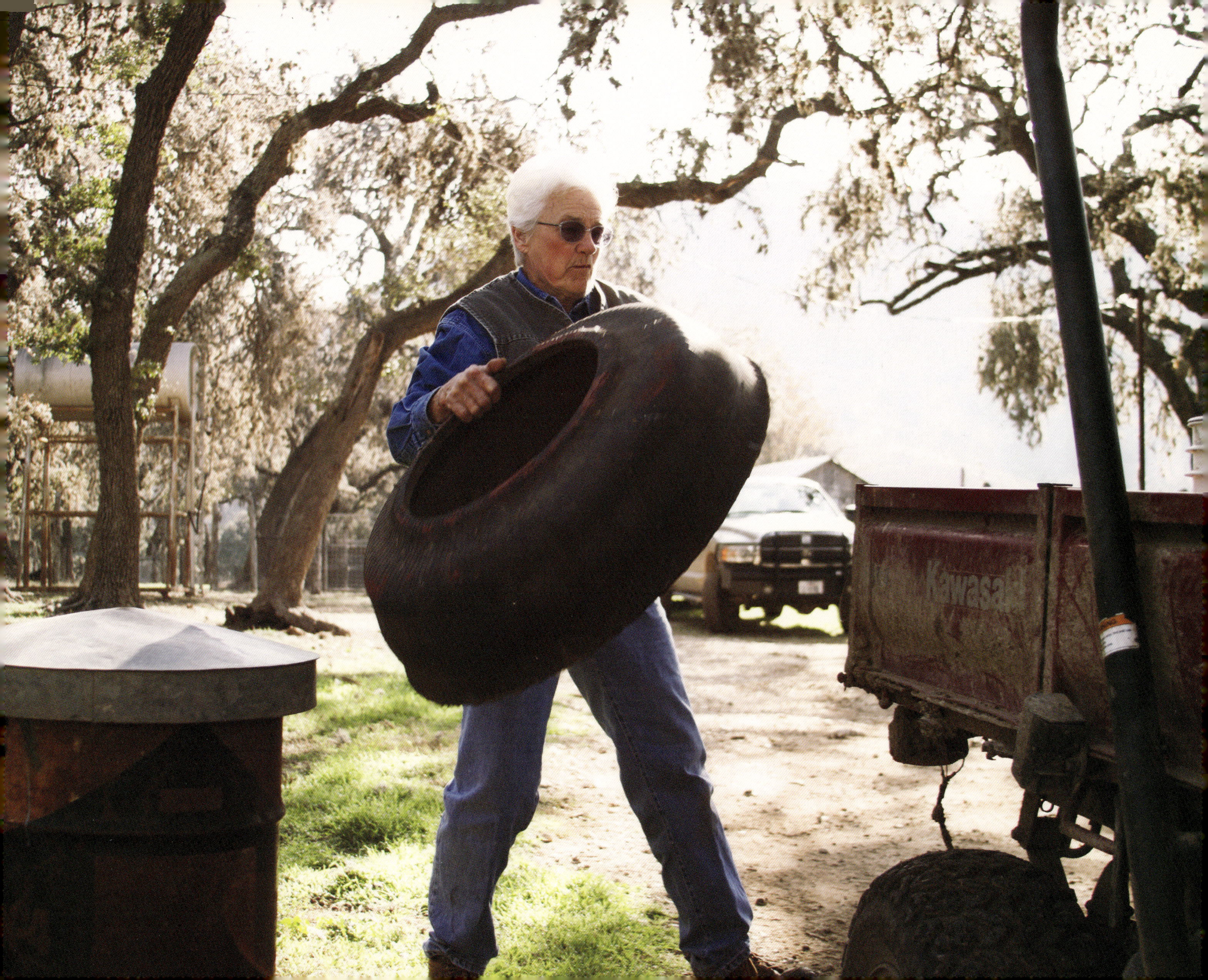
Kawasaki

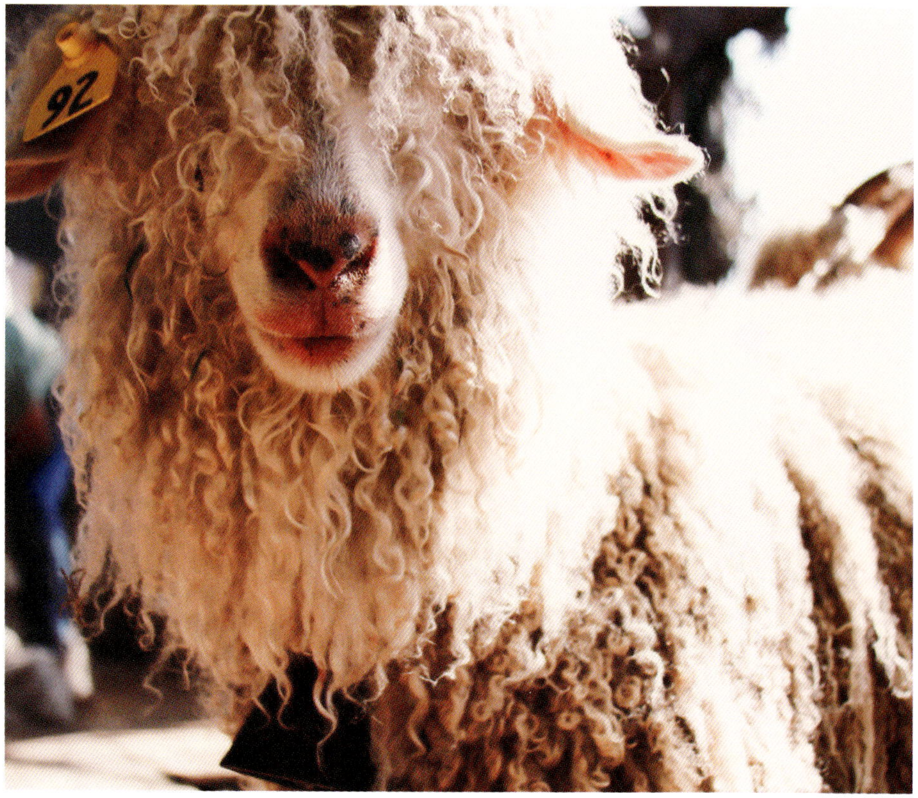
92

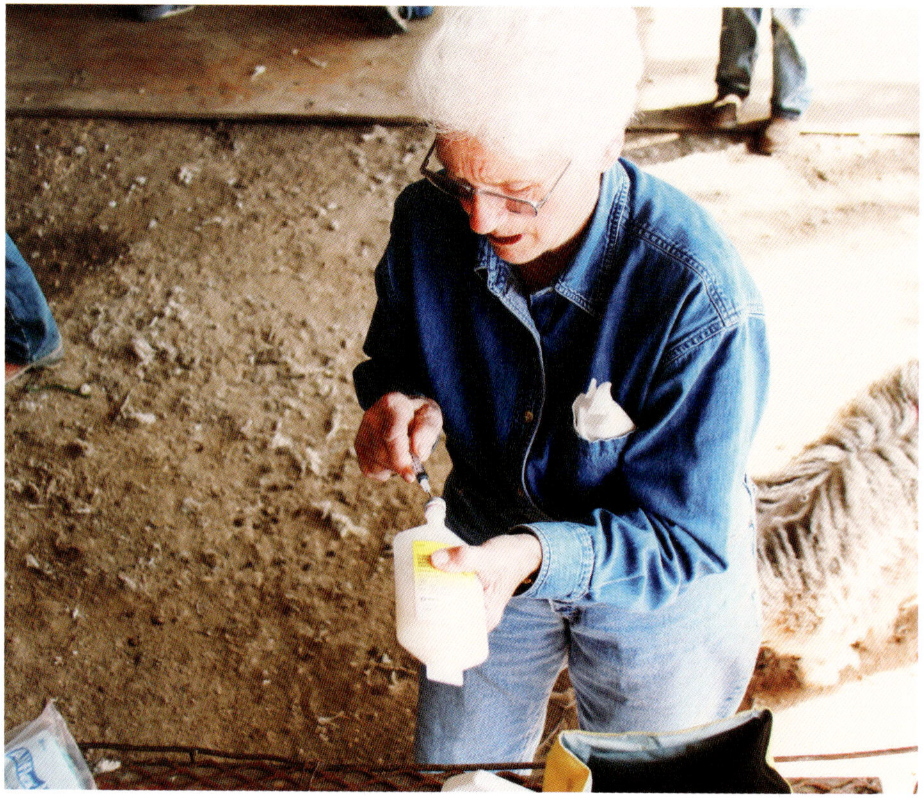

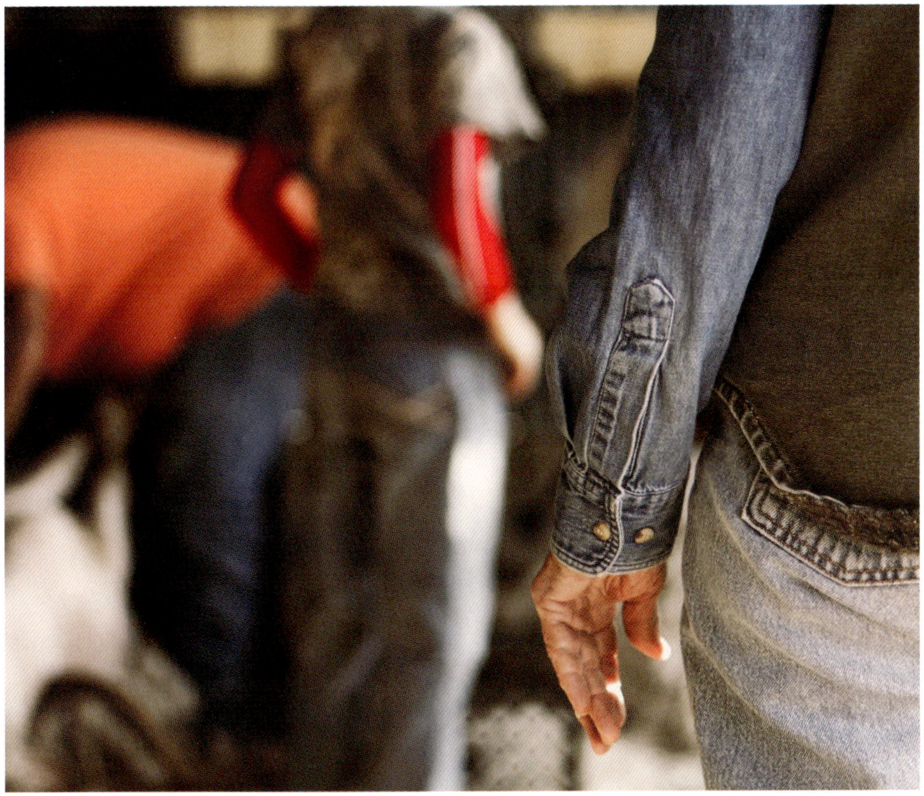

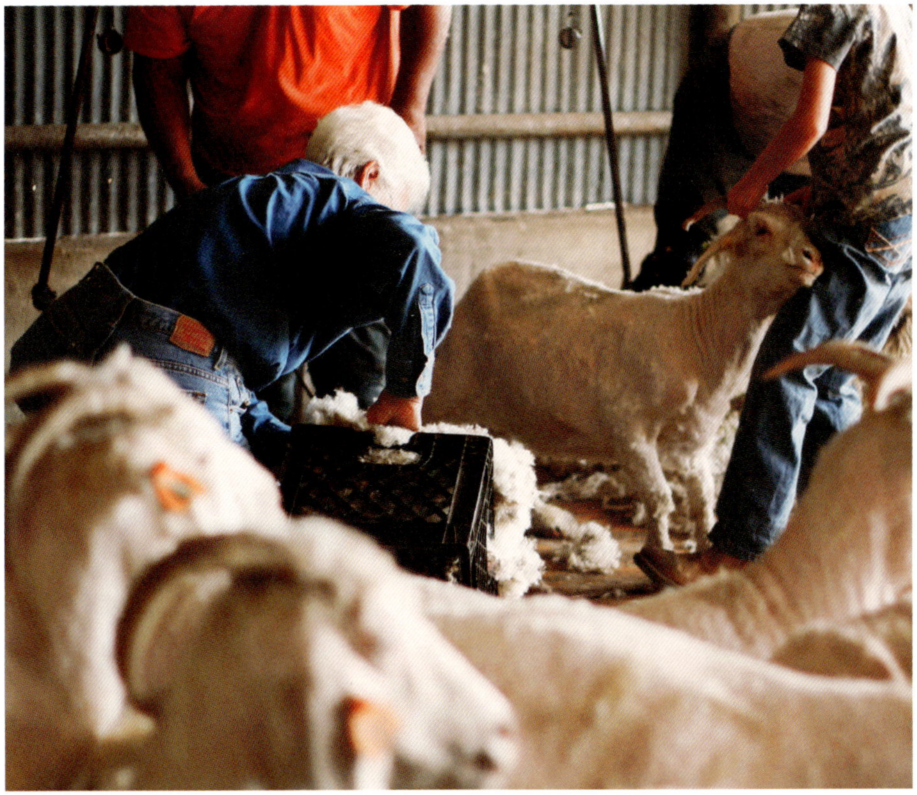

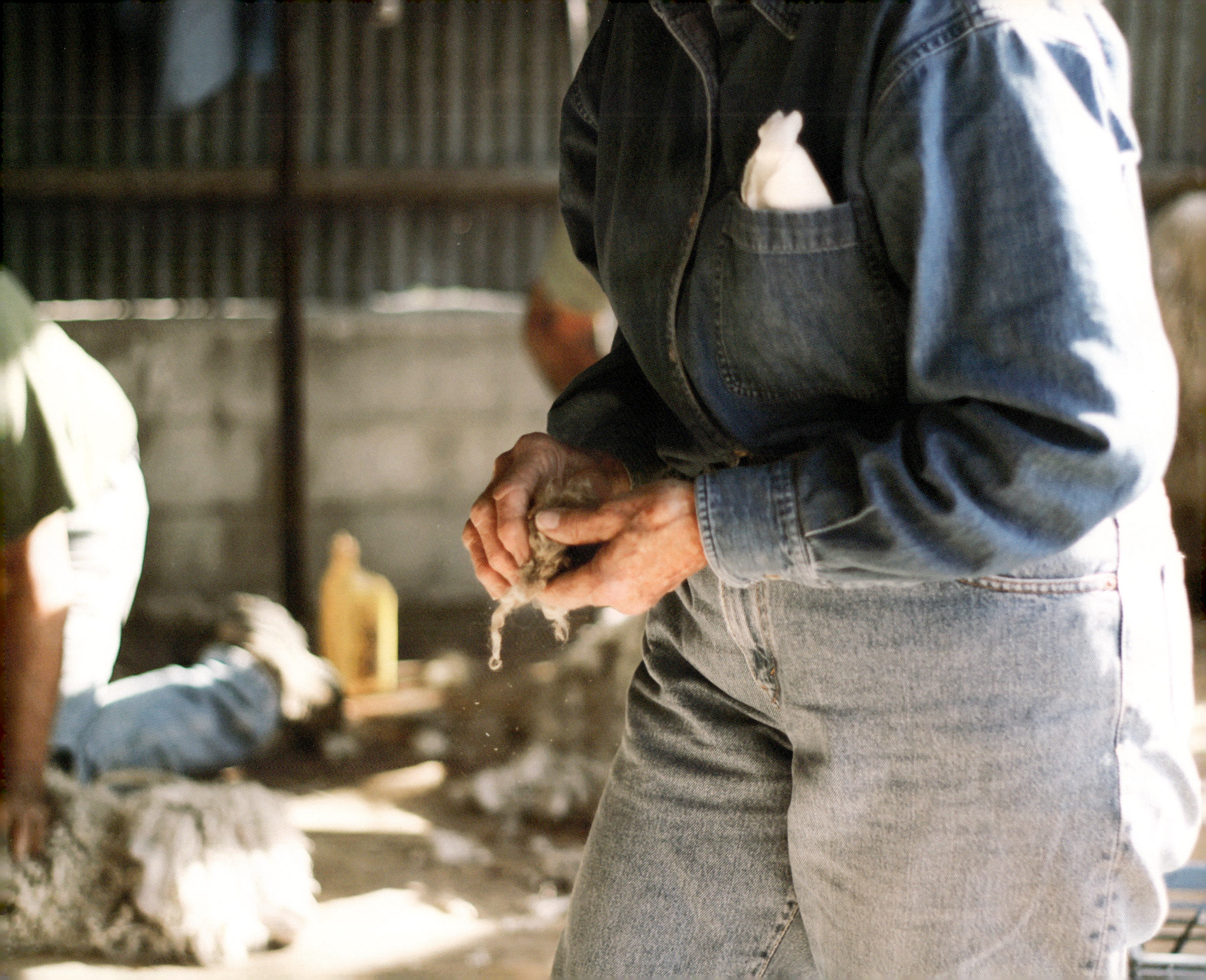

Keep'n On
< PLEASE NOTE FEEDING INSTRUCTIONS >
KID
MILK
REPLACER

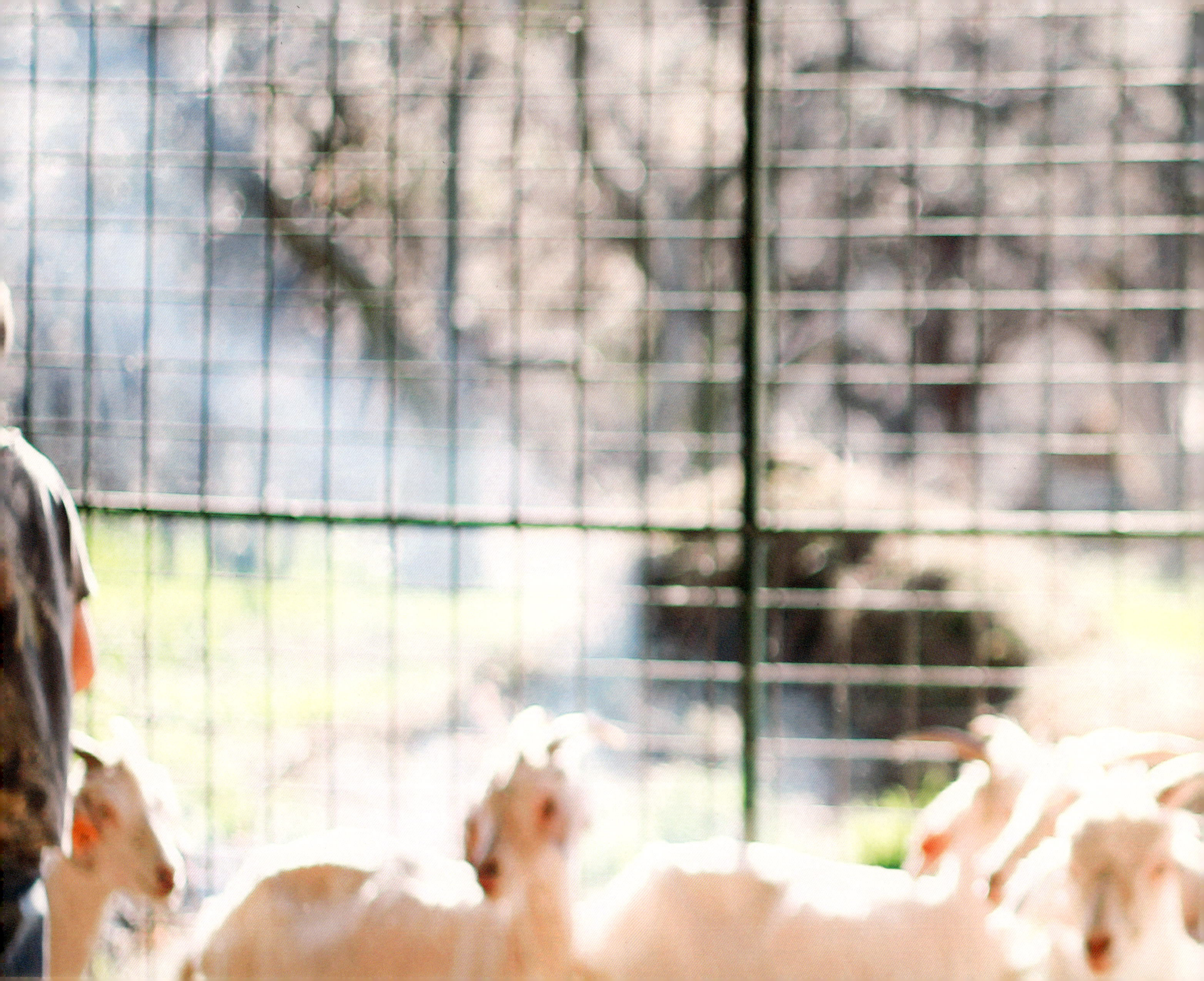

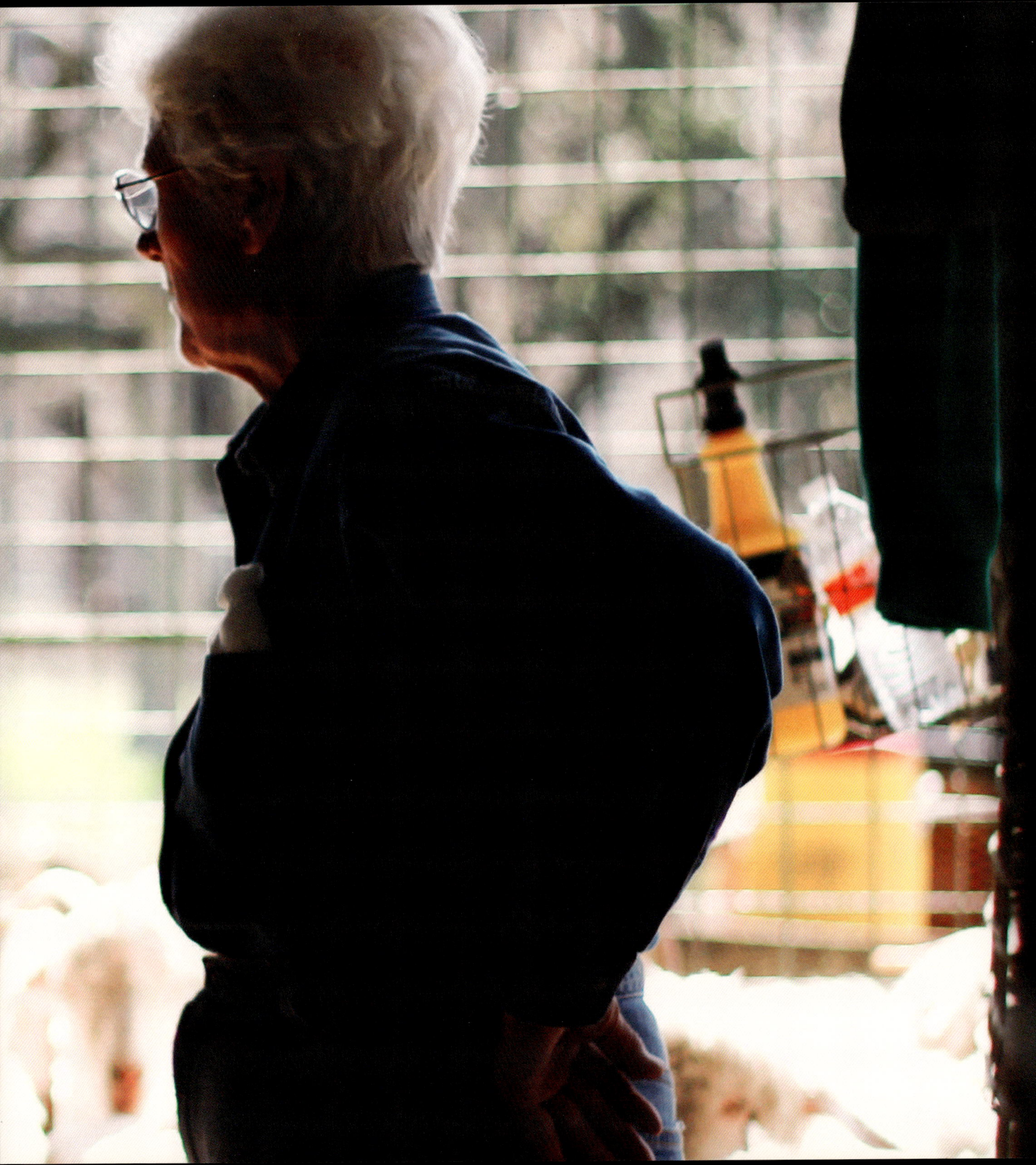

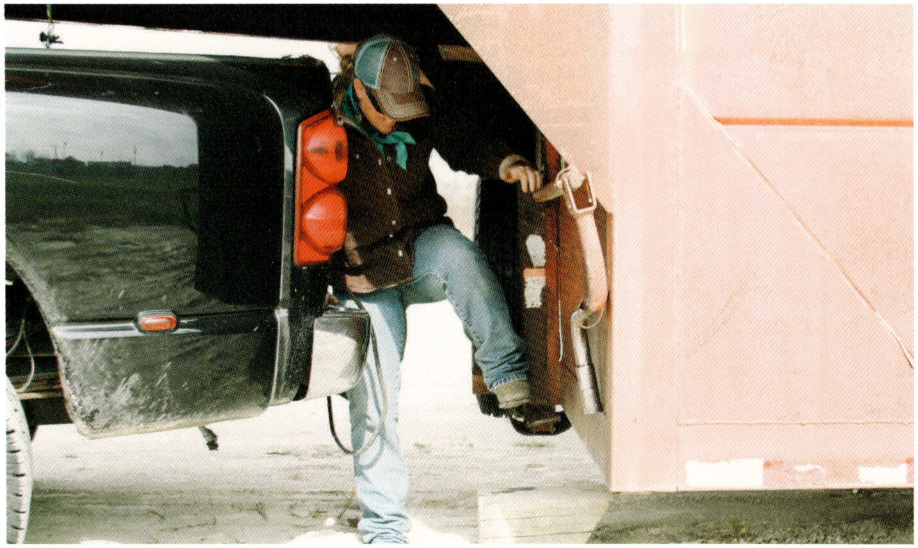

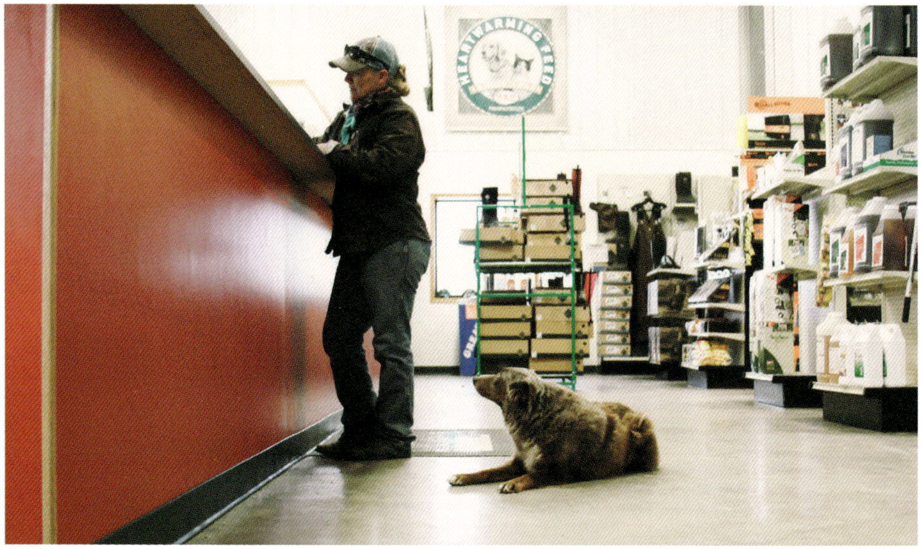

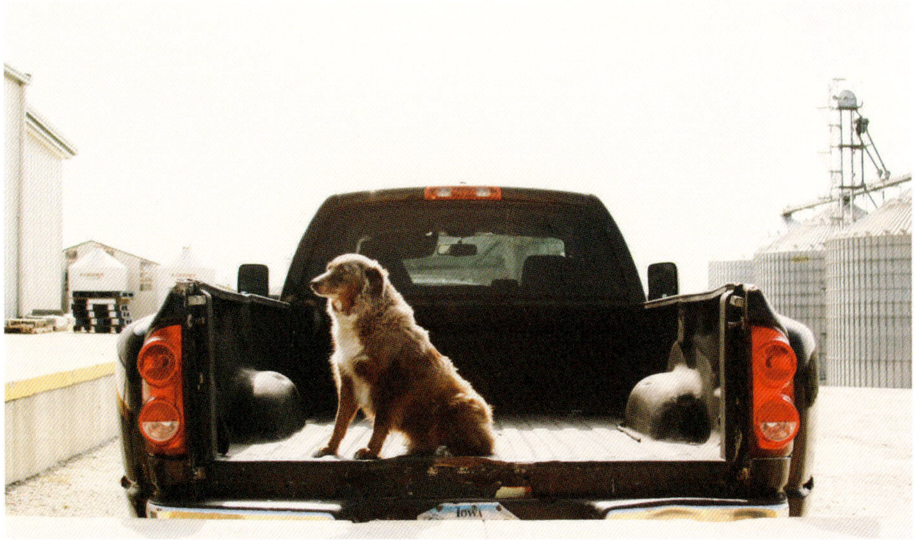

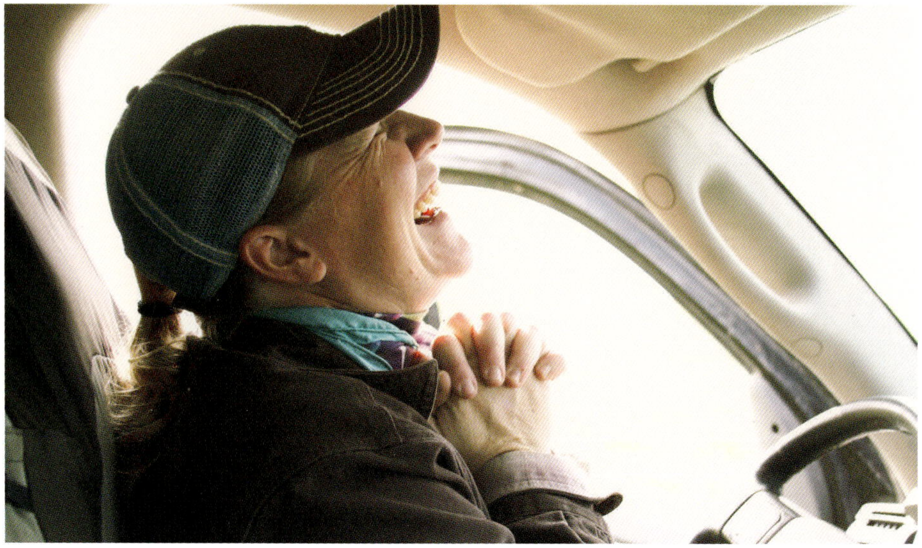

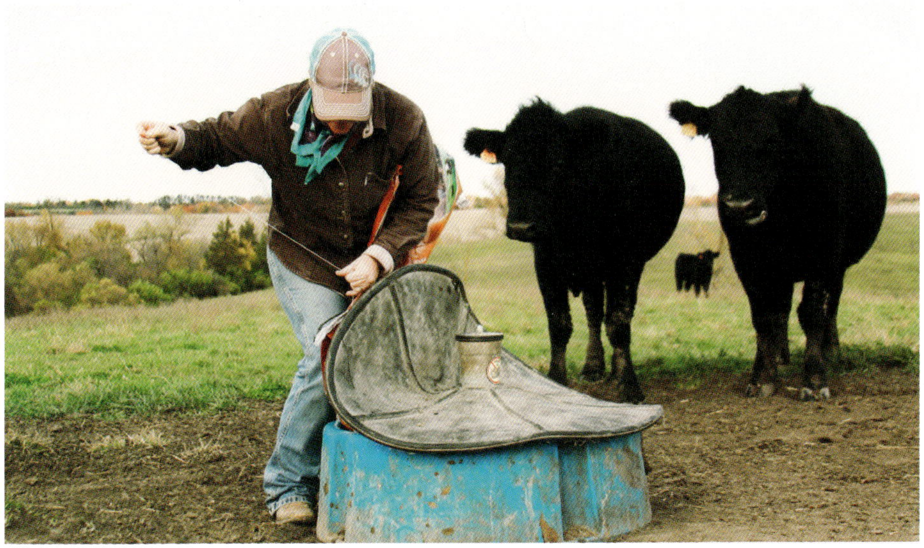

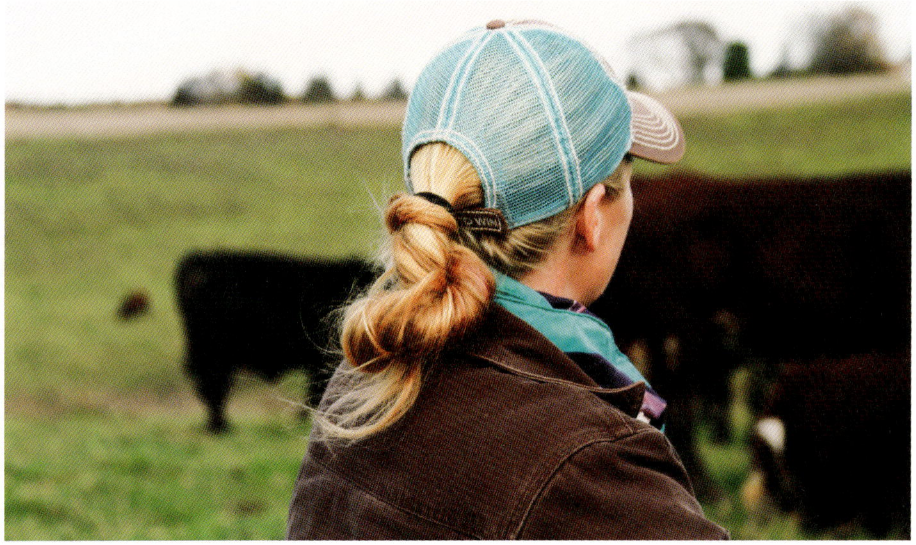

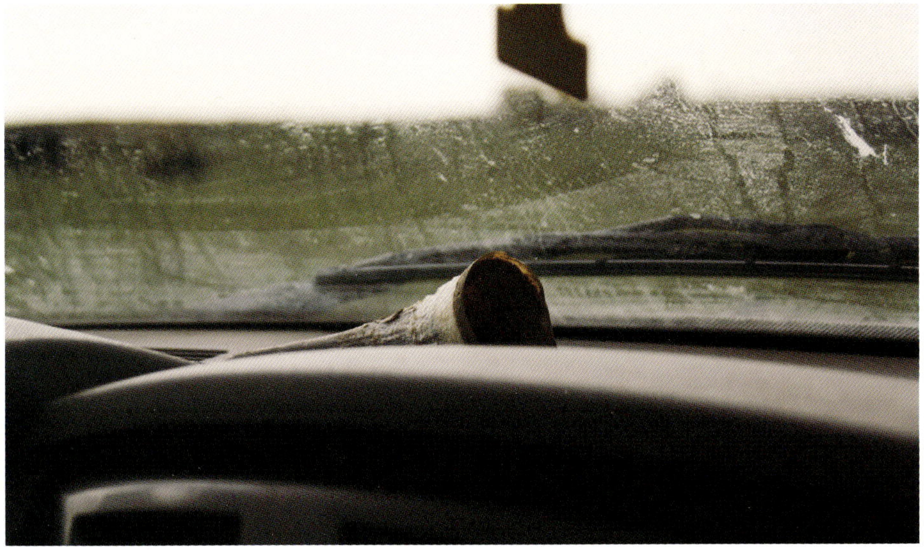

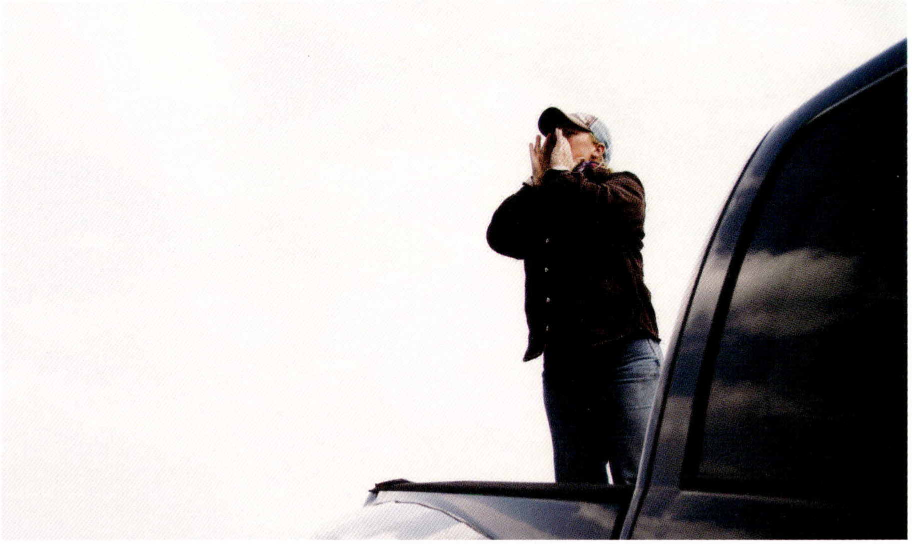

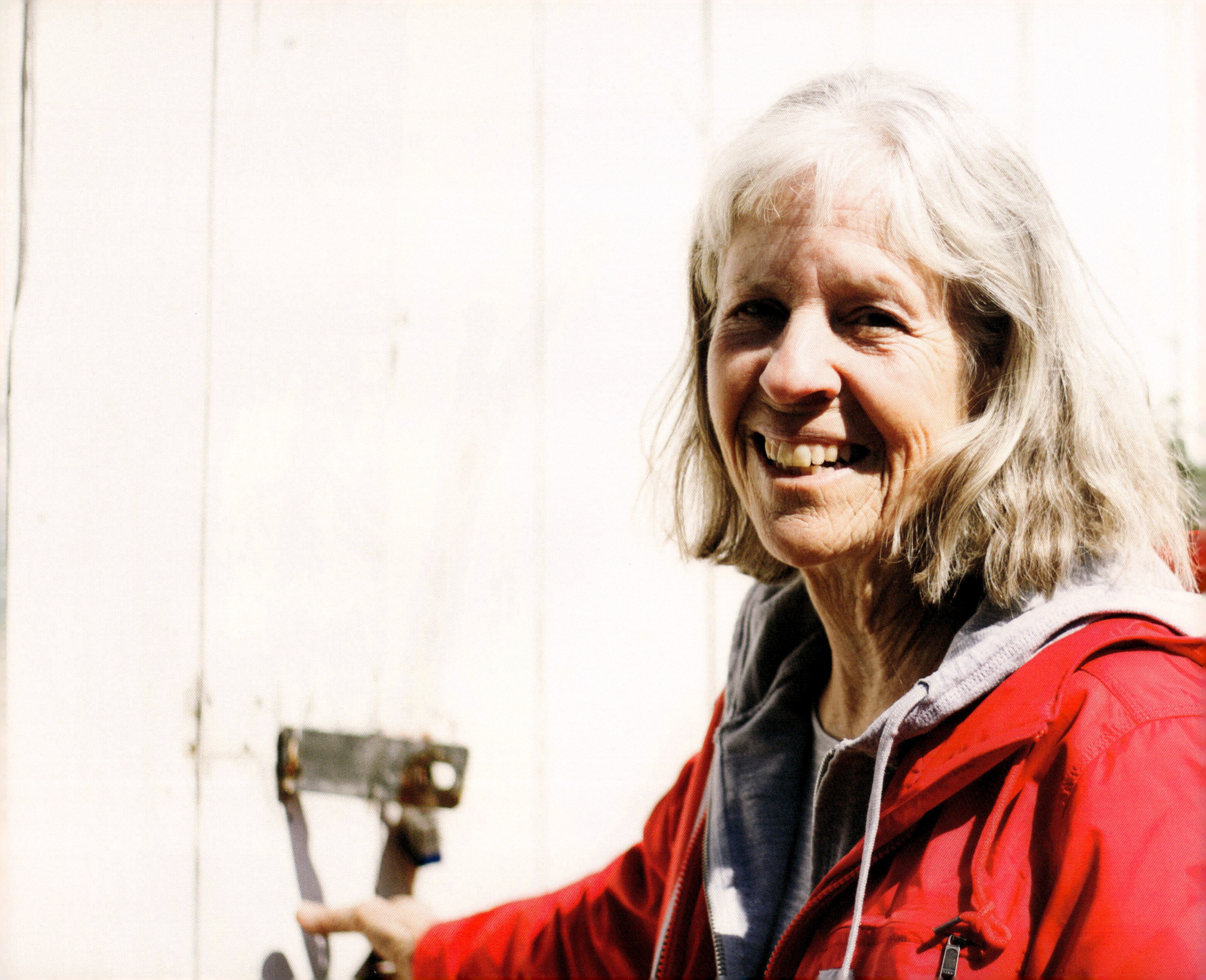

CAROL ANN
BOGGY CREEK FARM • TEXAS

Carol Ann Sayle created Boggy Creek with her husband in 1992. It was a USDA-certified organic urban farm before it was trendy to be one.

She farms two acres in Austin, Texas, and has more acres at another farm outside the city. The location in East Austin is one of the longest running urban farms in the United States and boasts an on-farm produce stand where all the crops are sold to a steady stream of regular customers seeking fresh, local food.

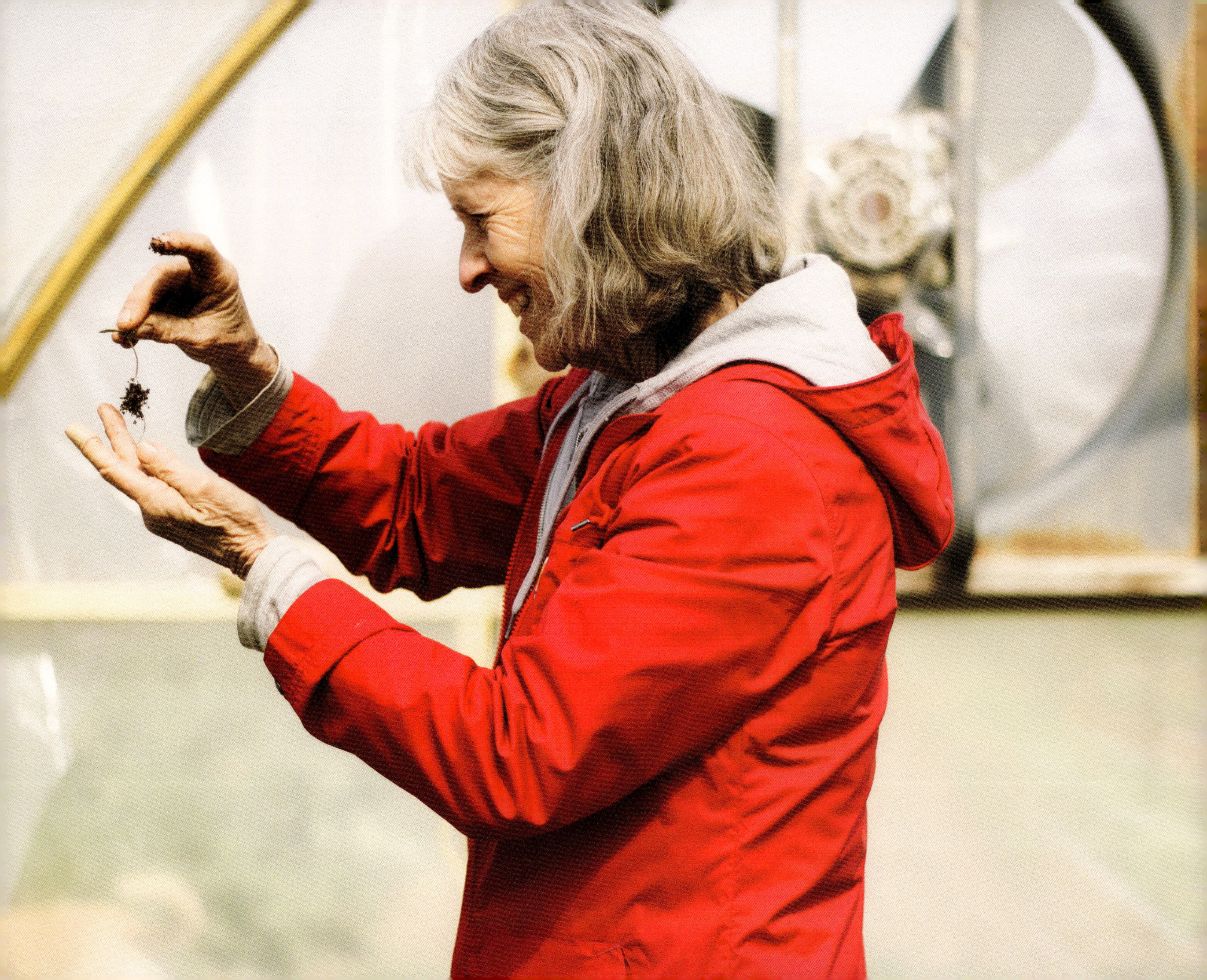

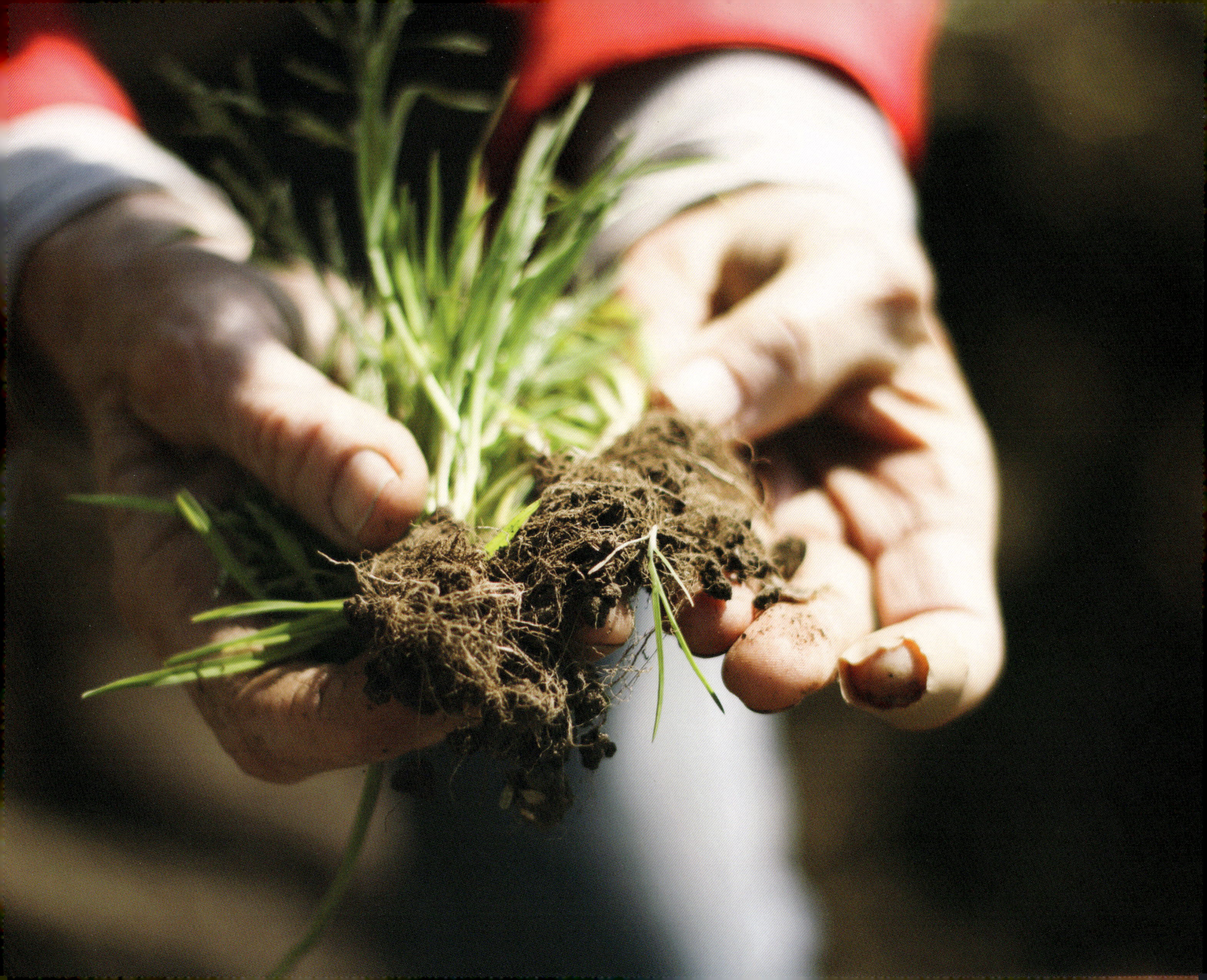

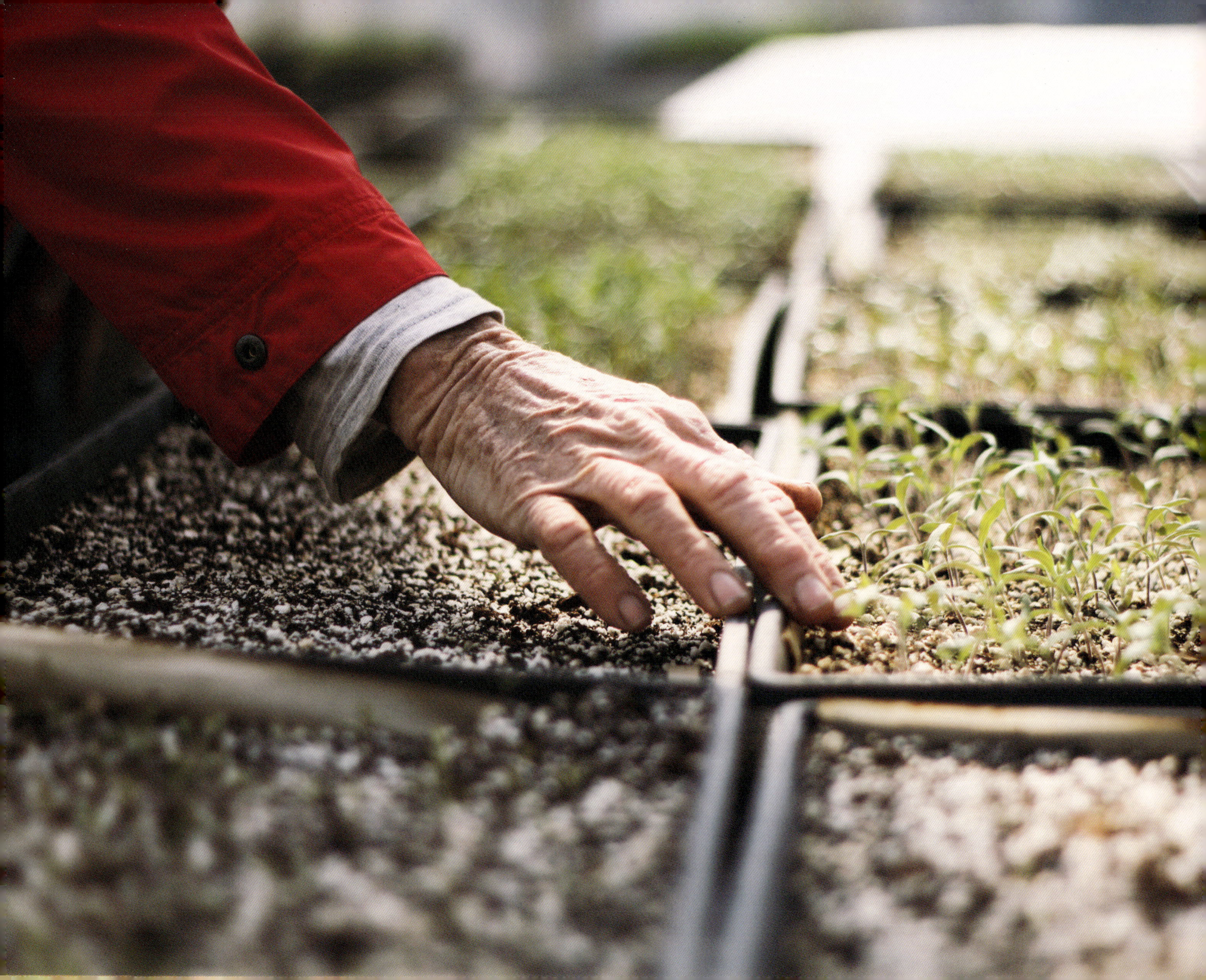

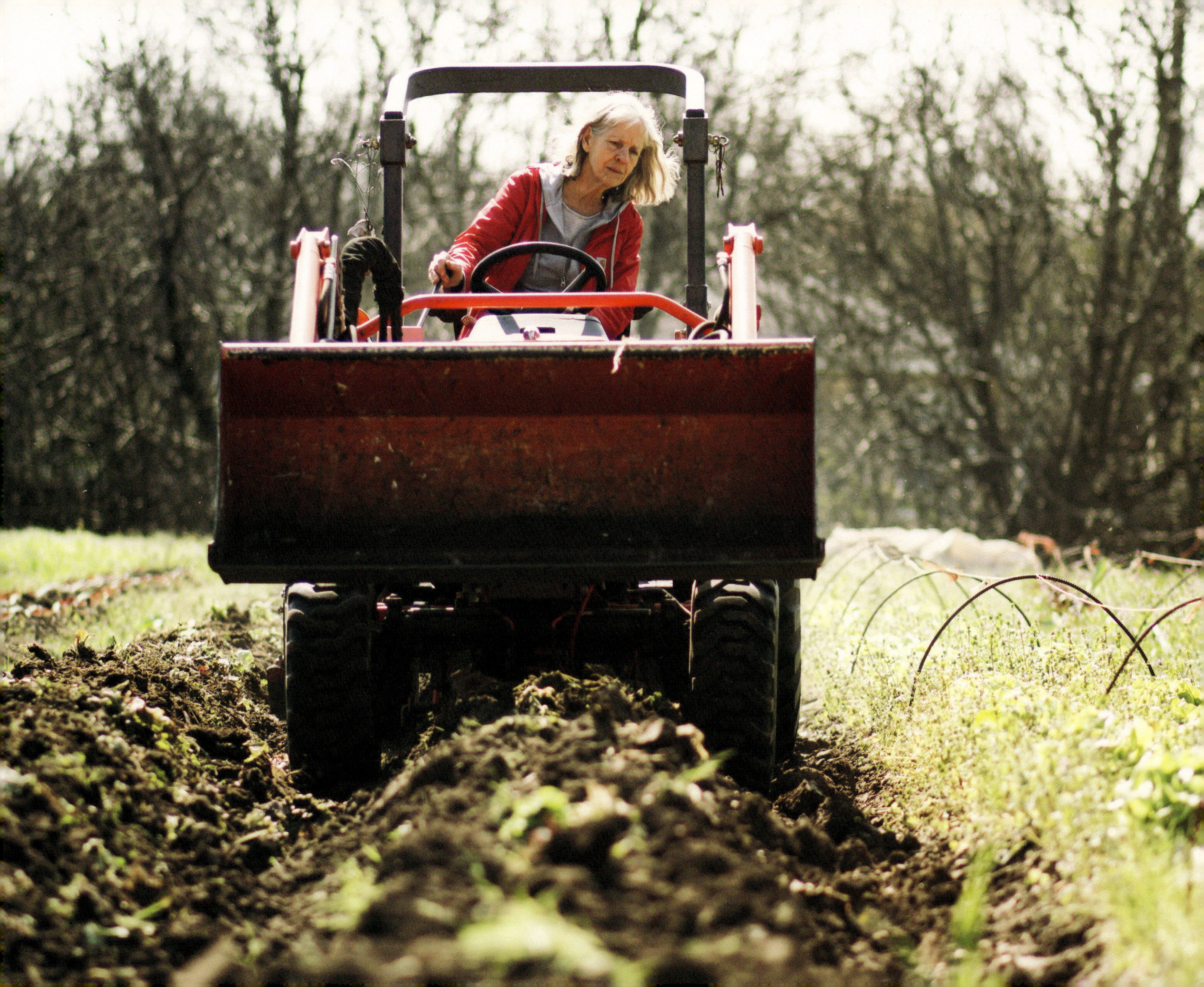

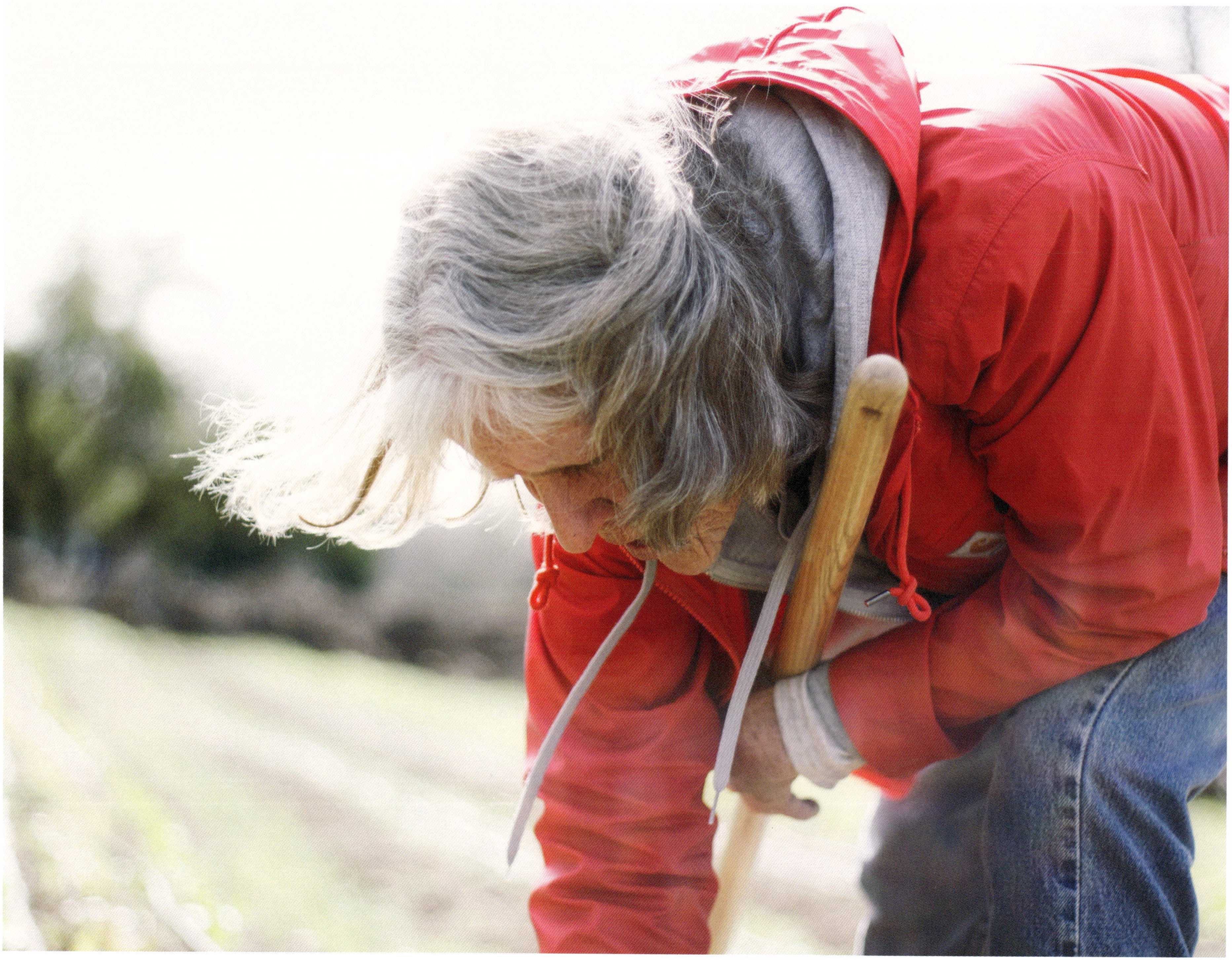

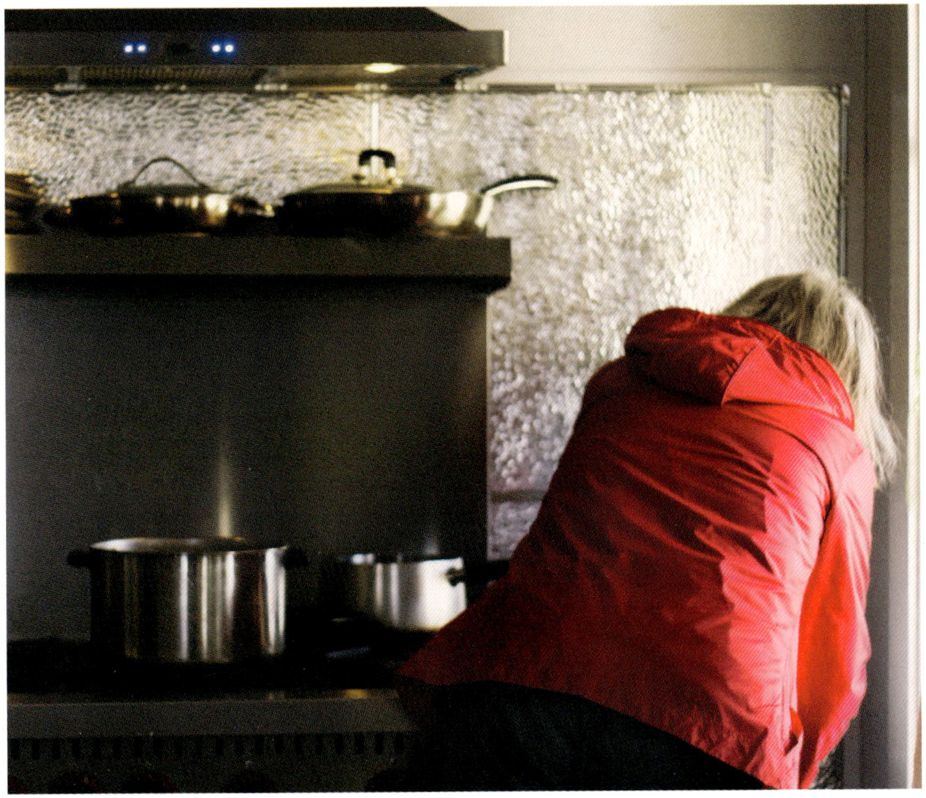

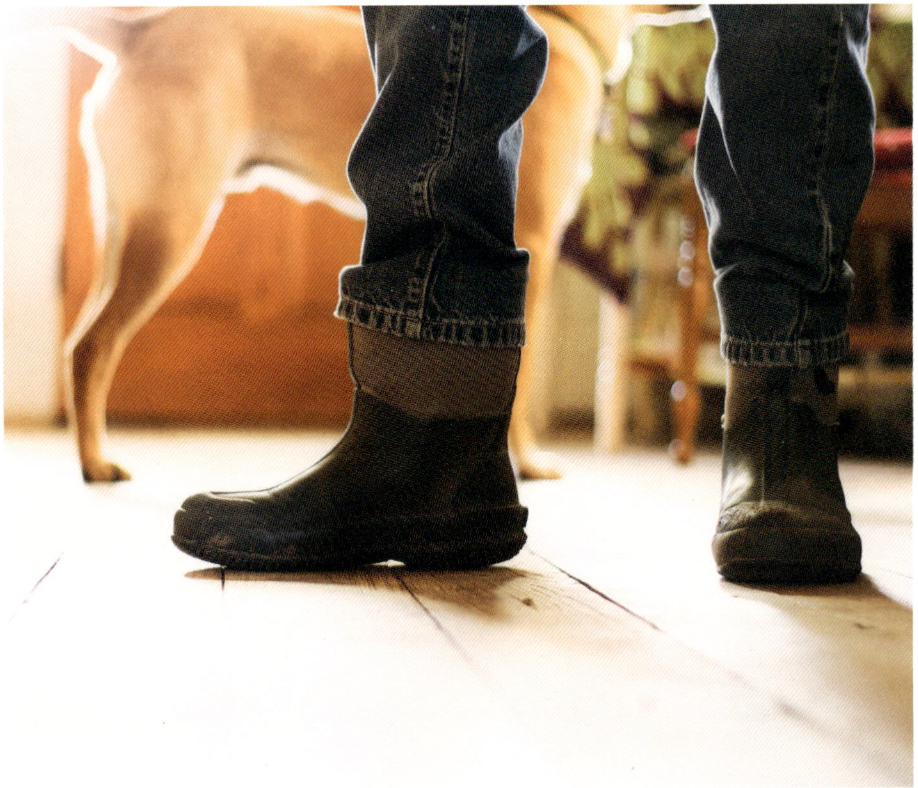

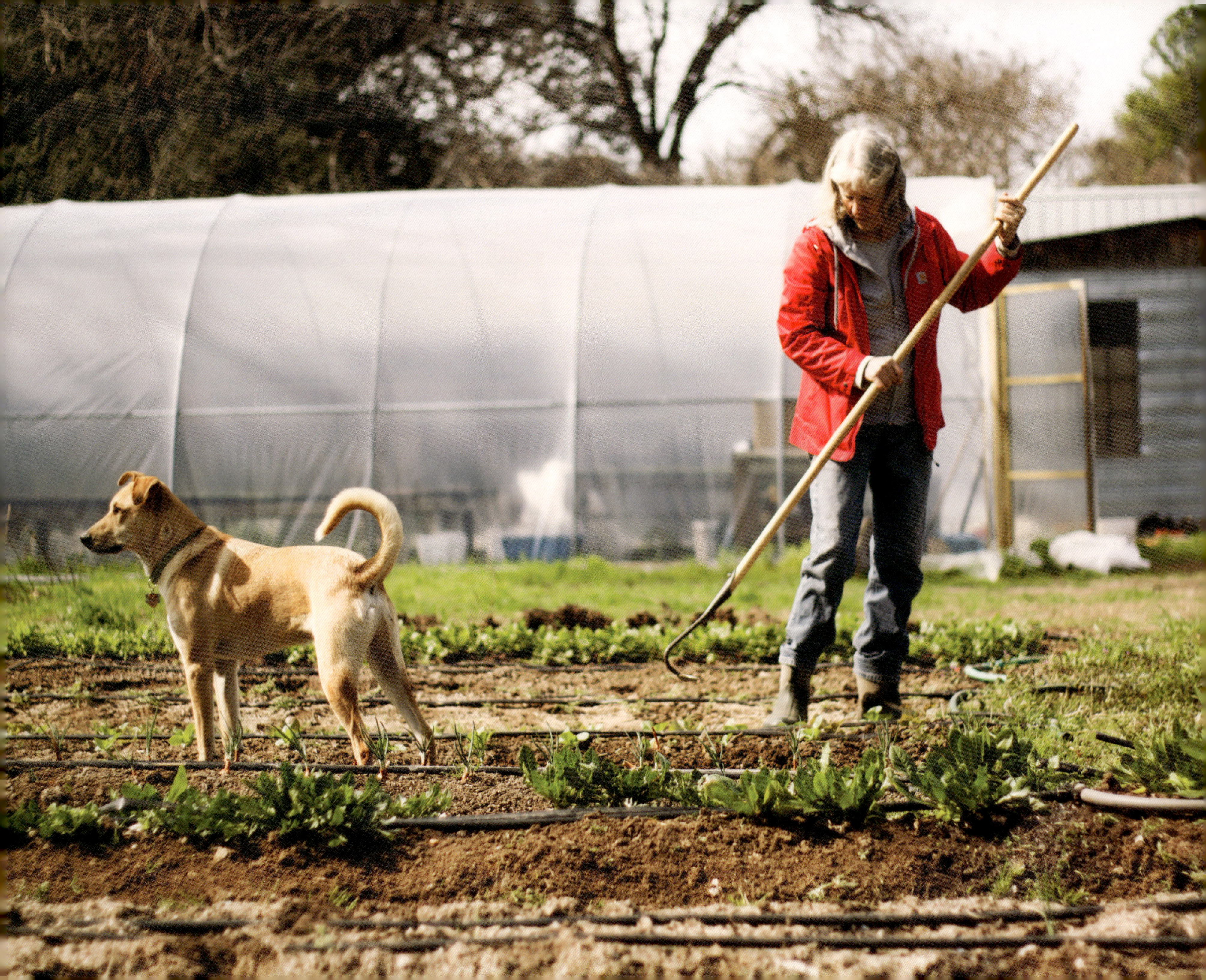

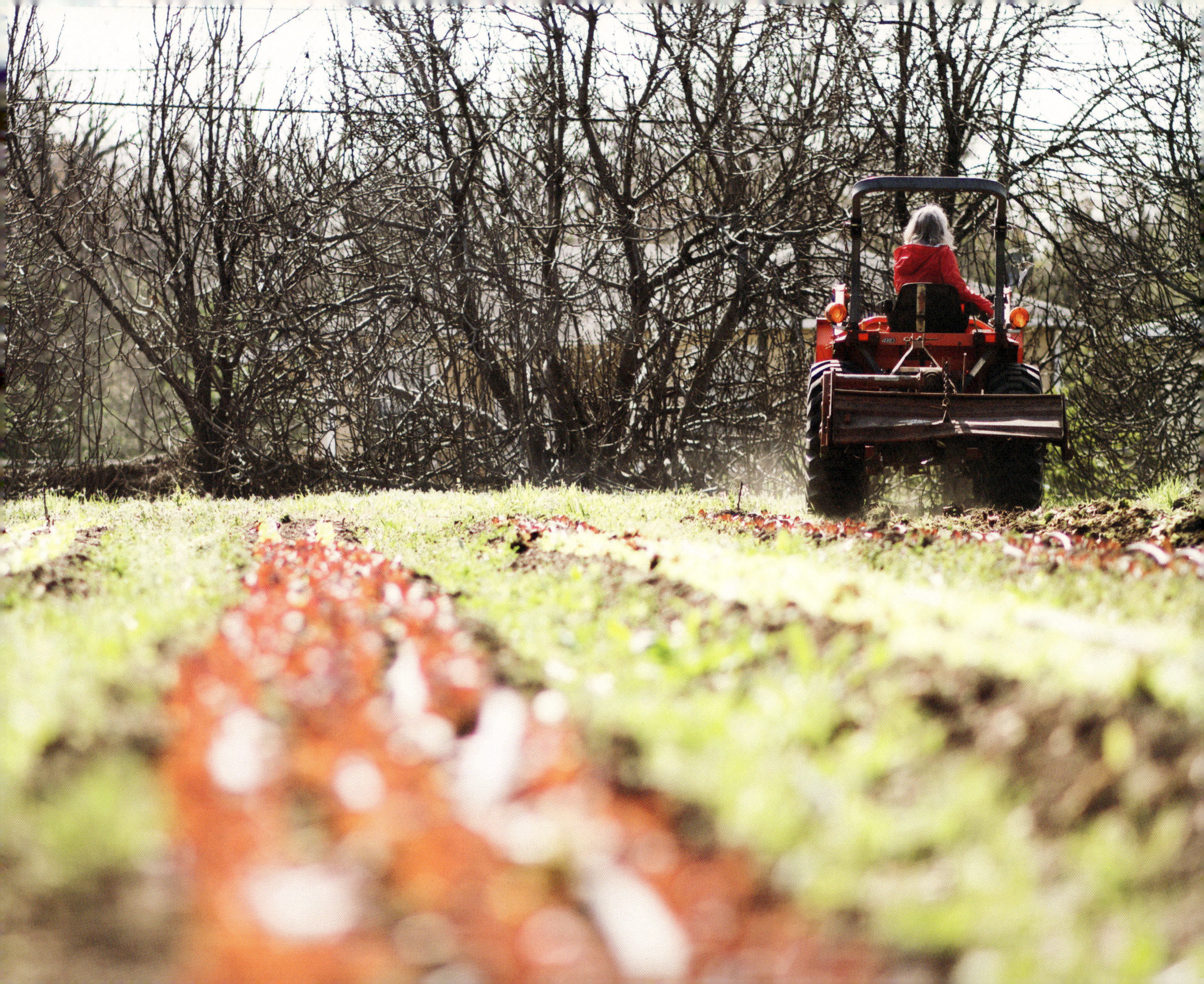

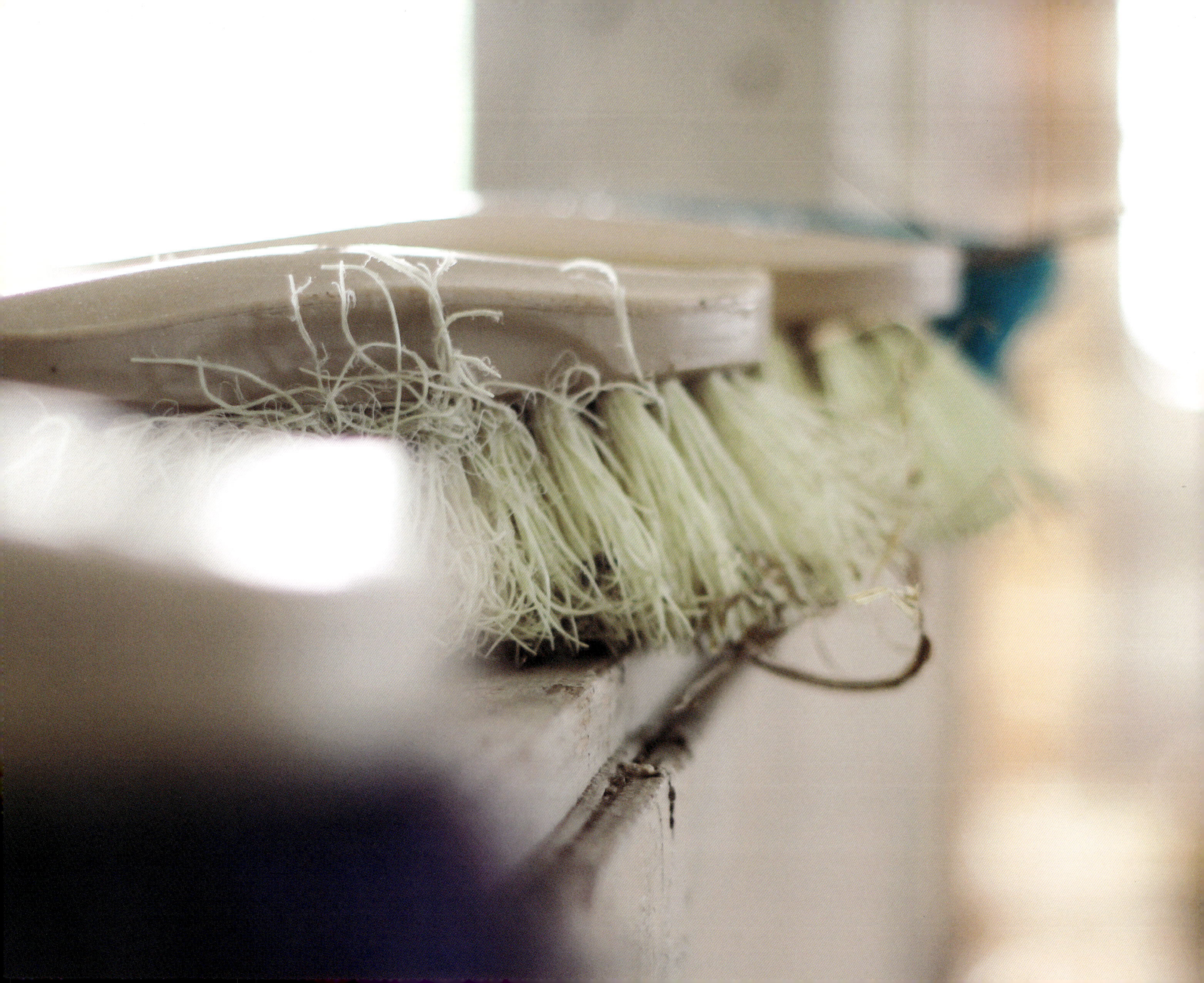

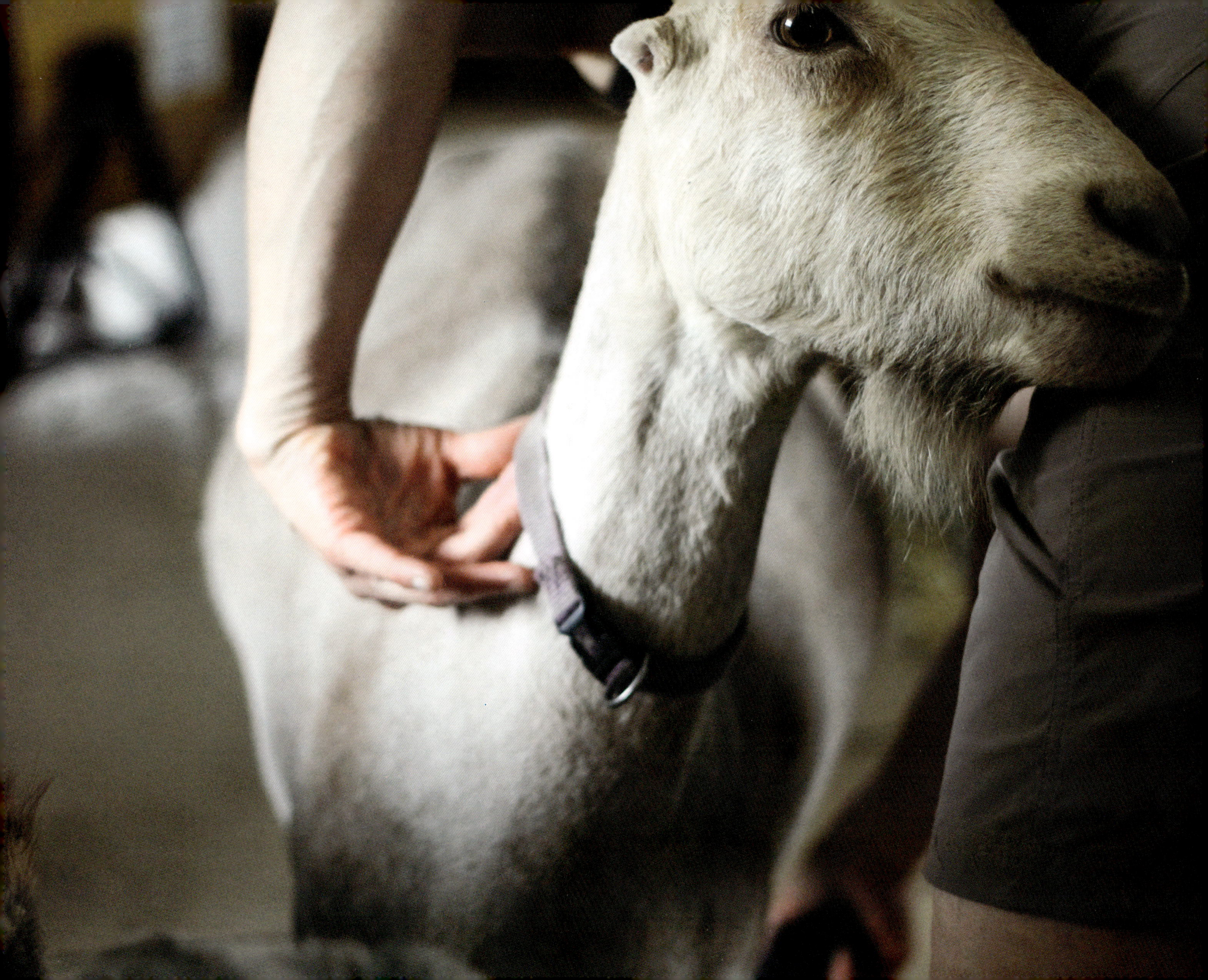

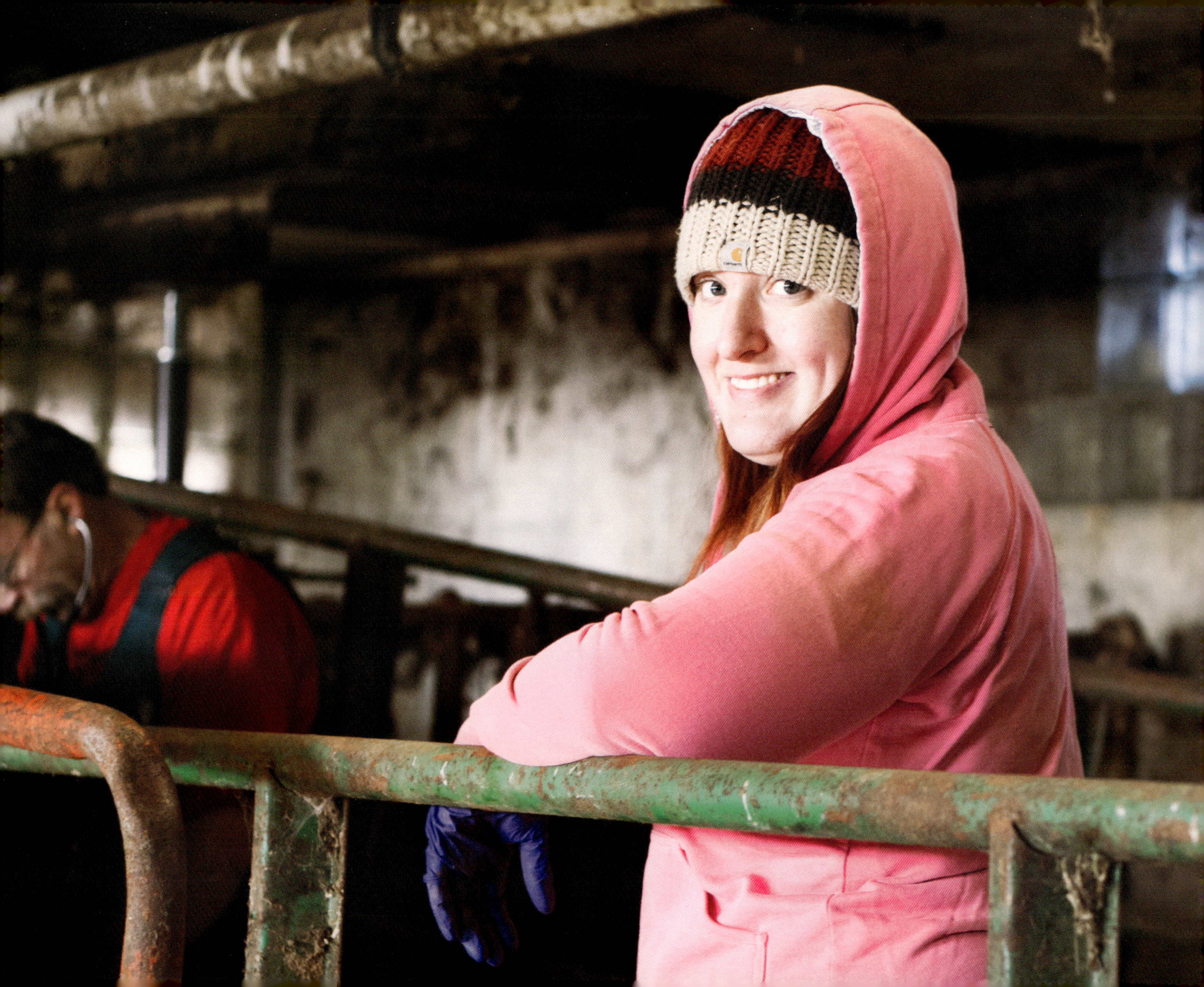

CARRIE
MESA DAIRY • WISCONSIN

Carrie Mess works alongside her husband Patrick and his parents on their family-owned and operated dairy in Wisconsin. Though she wasn't raised in agriculture, she has always been an animal lover and quickly fell for the cattle when she got married.

Carrie is also a bit of a star in the social media world with her blog The Adventures of Dairy Carrie.

Carrie's family experienced tragedy last year when her mother-in-law was injured in a farming accident, which left the dairy operation down one person. Then, in August, Carrie became a mother to a baby boy. Silas goes everywhere with her as she tends to the dairy operation.

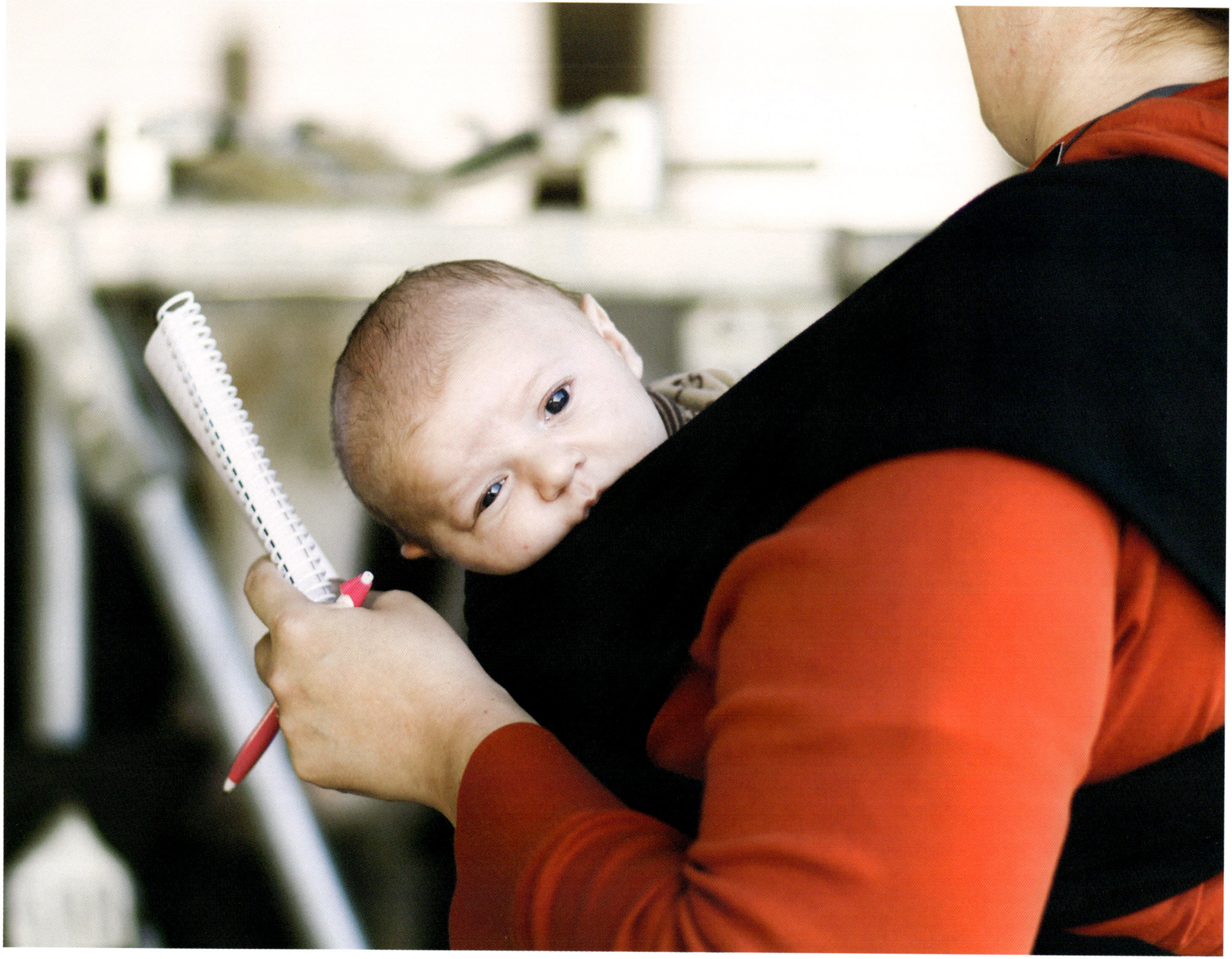

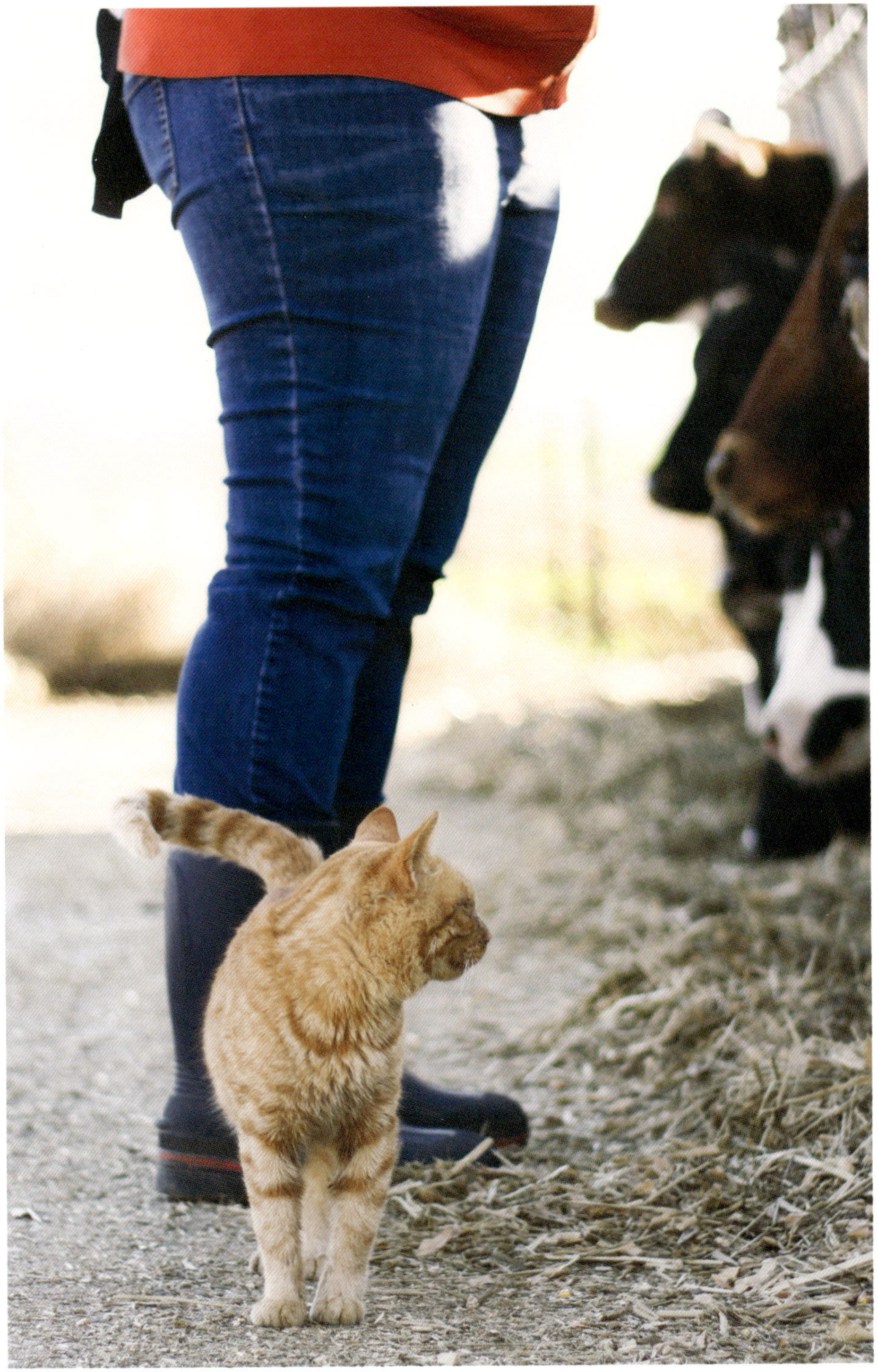

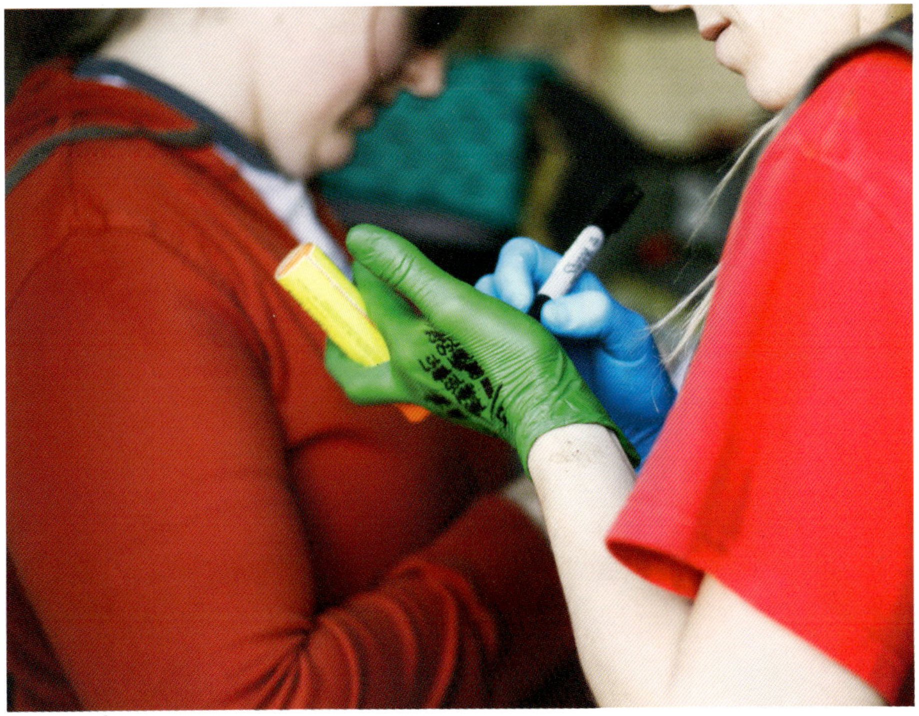

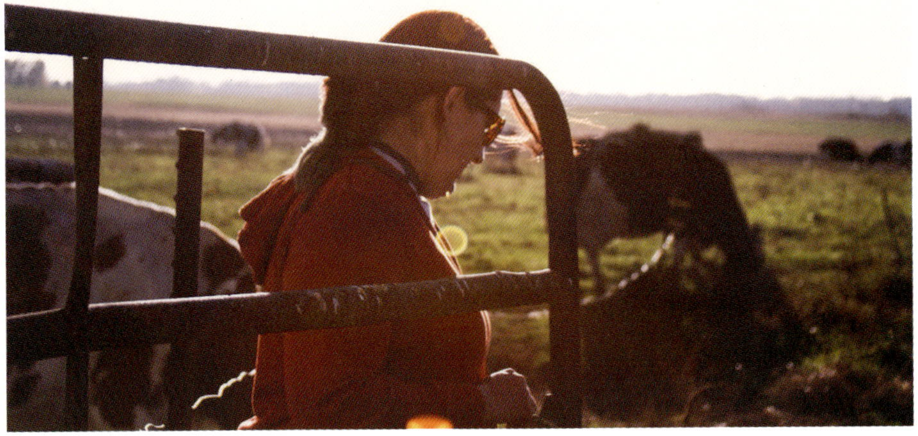

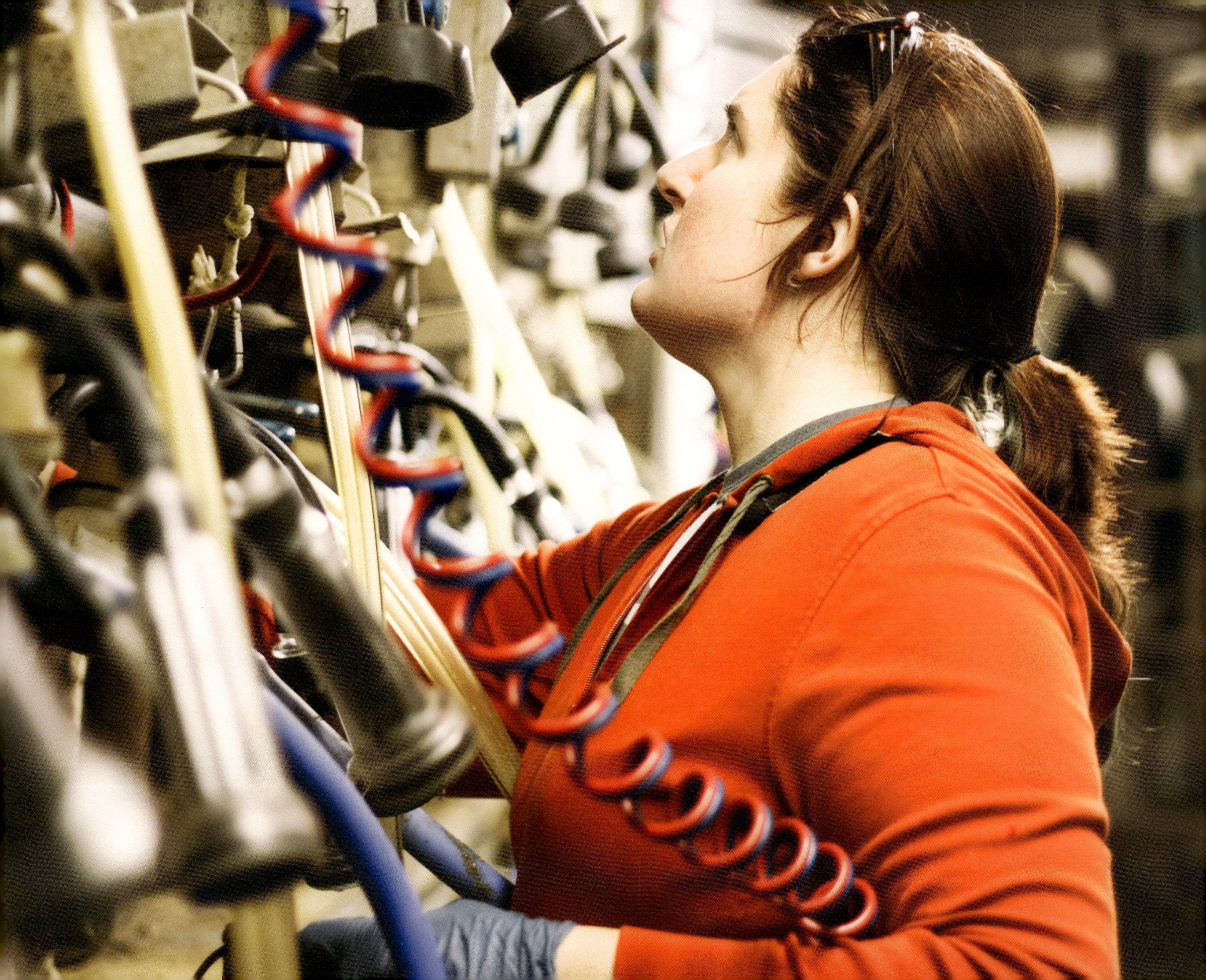

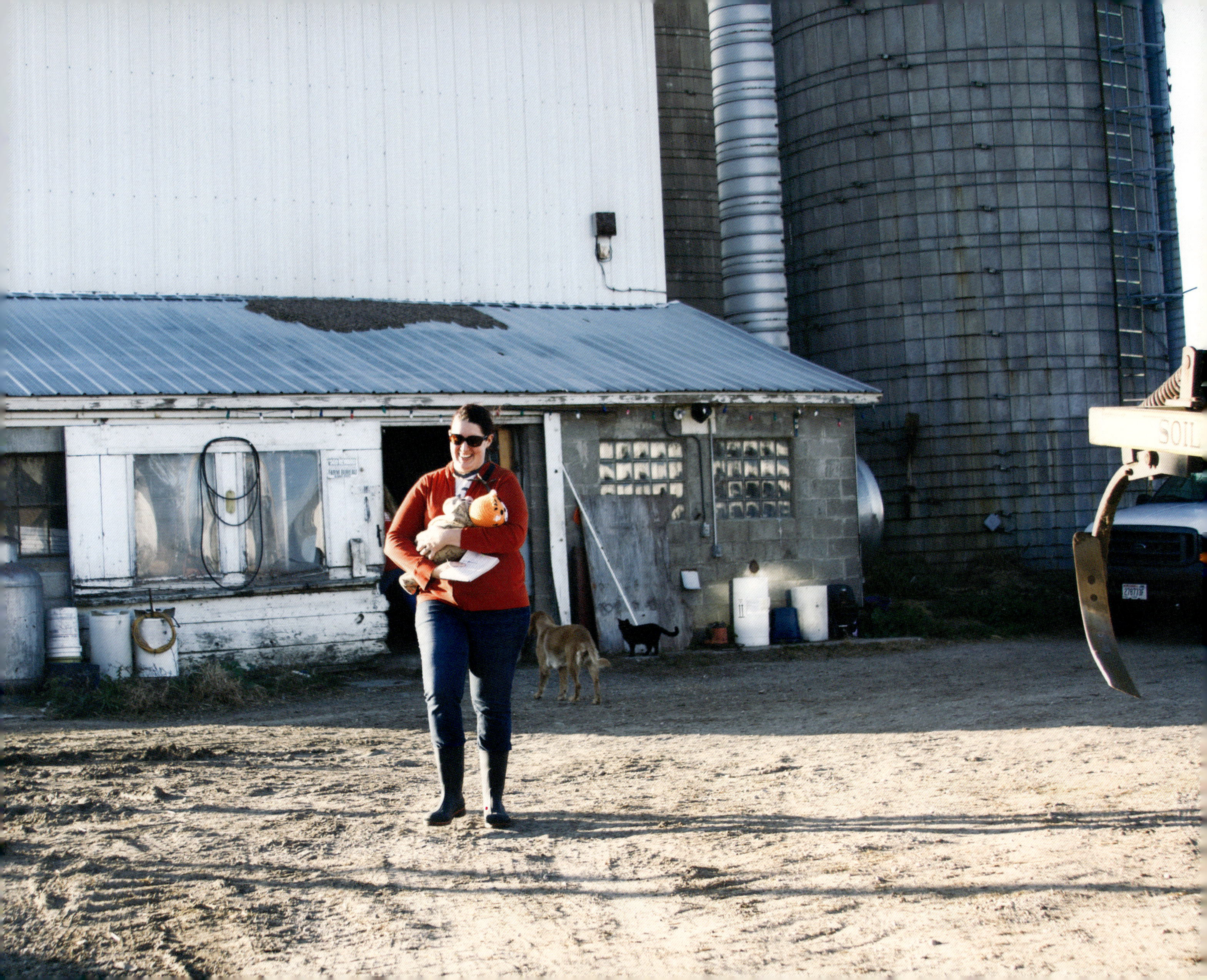
SOIL

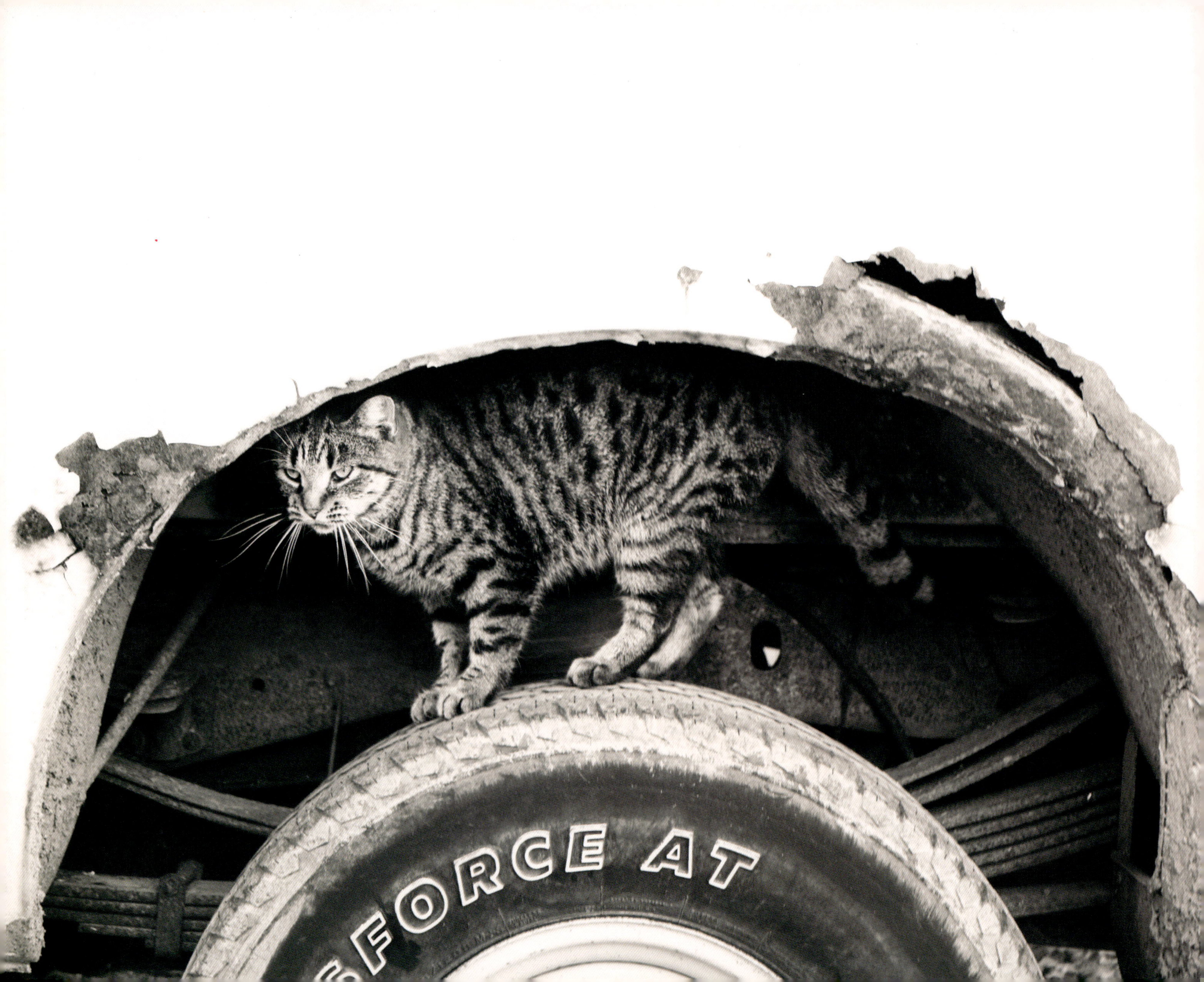
FORCE AT

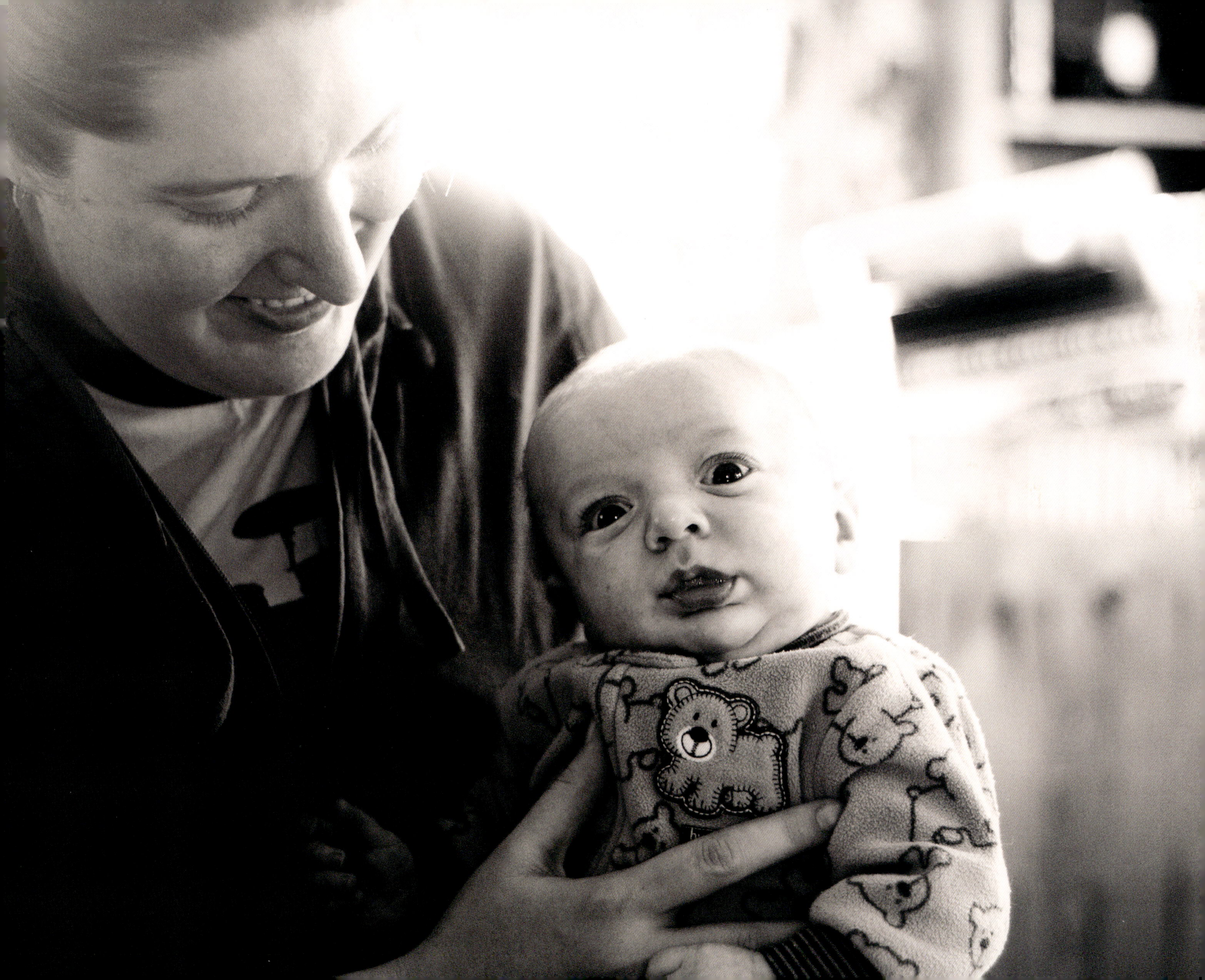

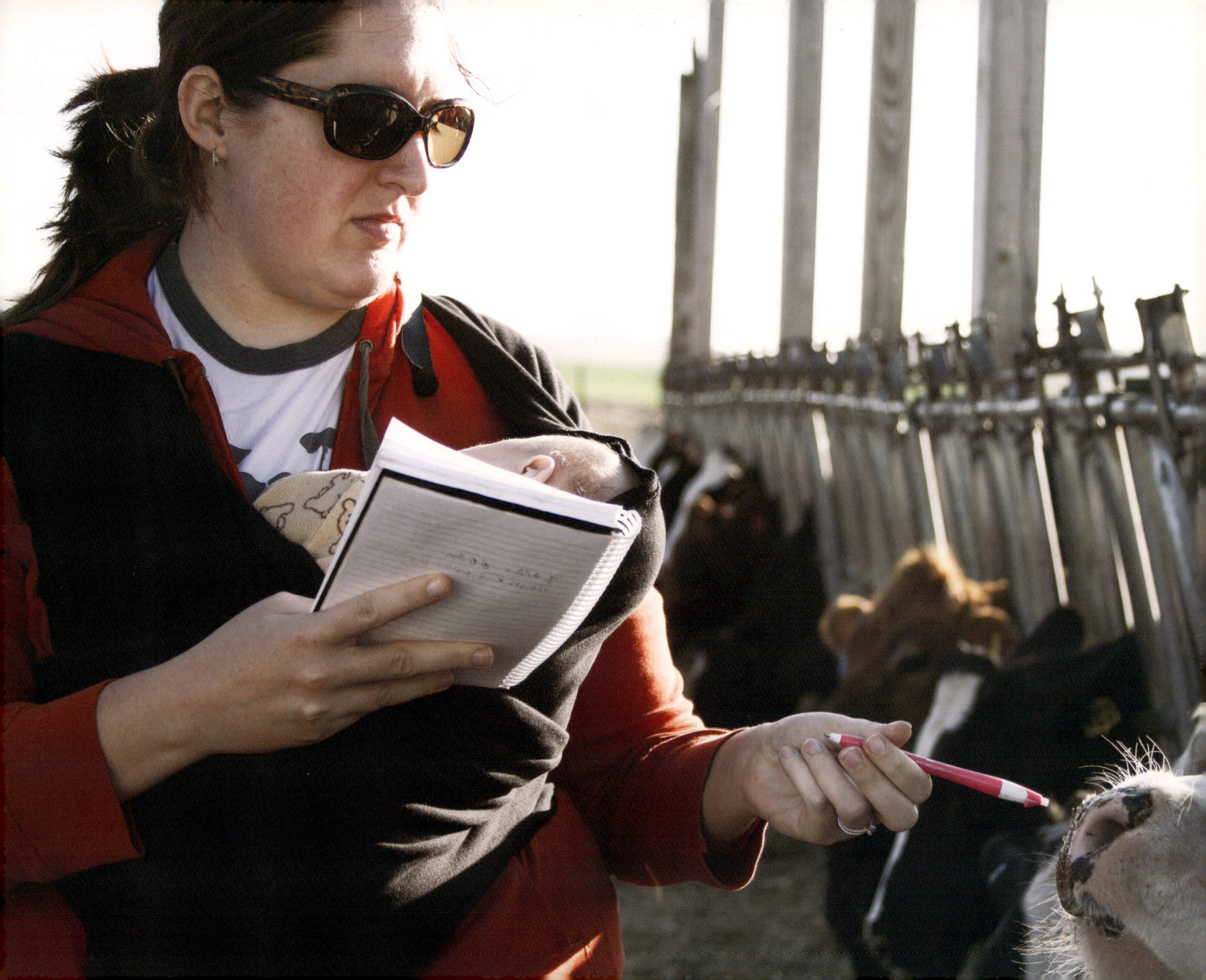

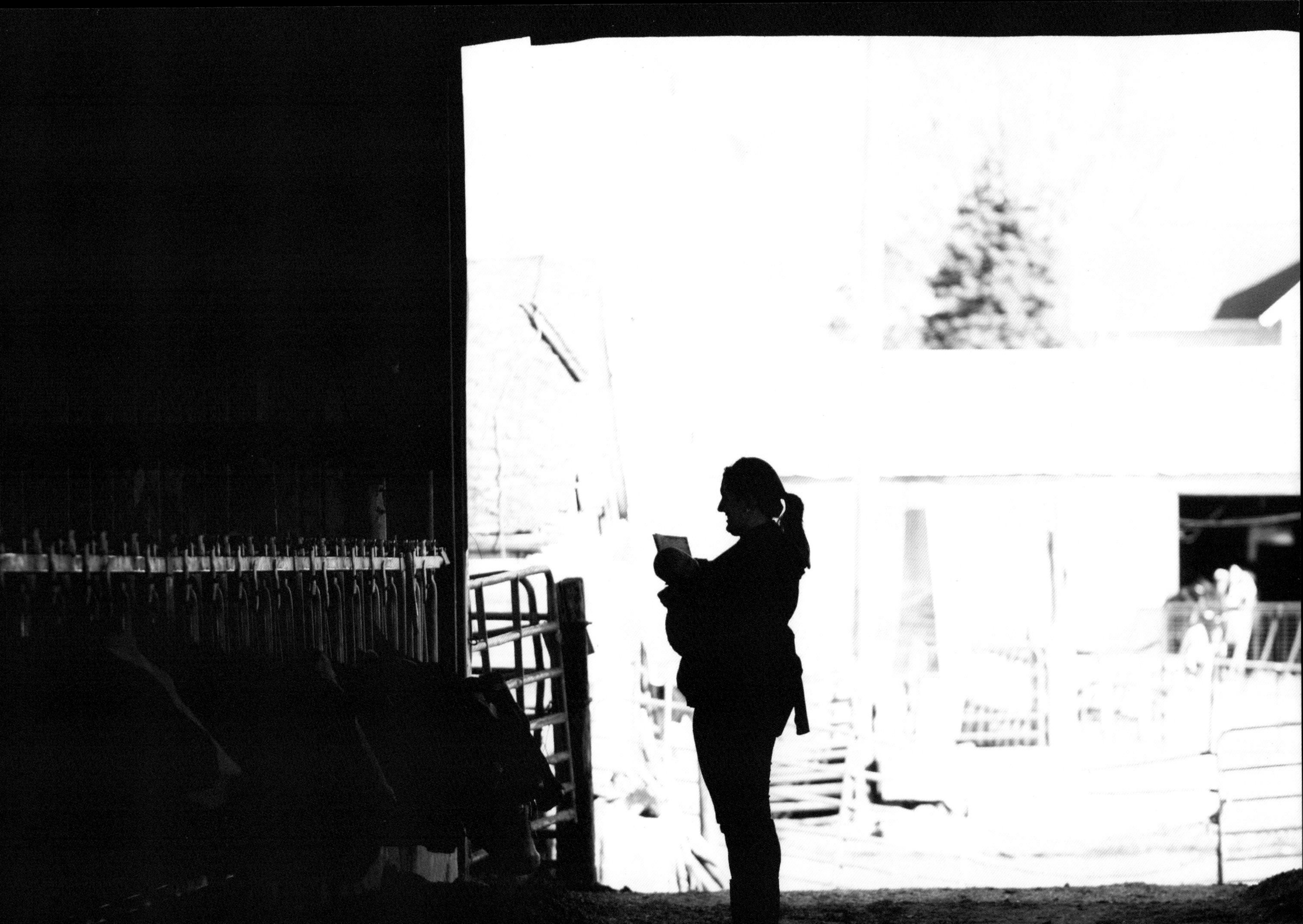

PAM

JOHNSON FAMILY FARM • IOWA

Pam Johnson raises corn and soybeans with her family near Floyd, Iowa. Her family is very important to her and she enjoys farming with her husband, two sons and their wives.

She long has been an advocate for agricultural research and rural economic development, and is a mentor for women in the industry. Pam has been involved with many organizations and was the first woman president of the National Corn Growers Association. She has also traveled the world advocating for rural America.

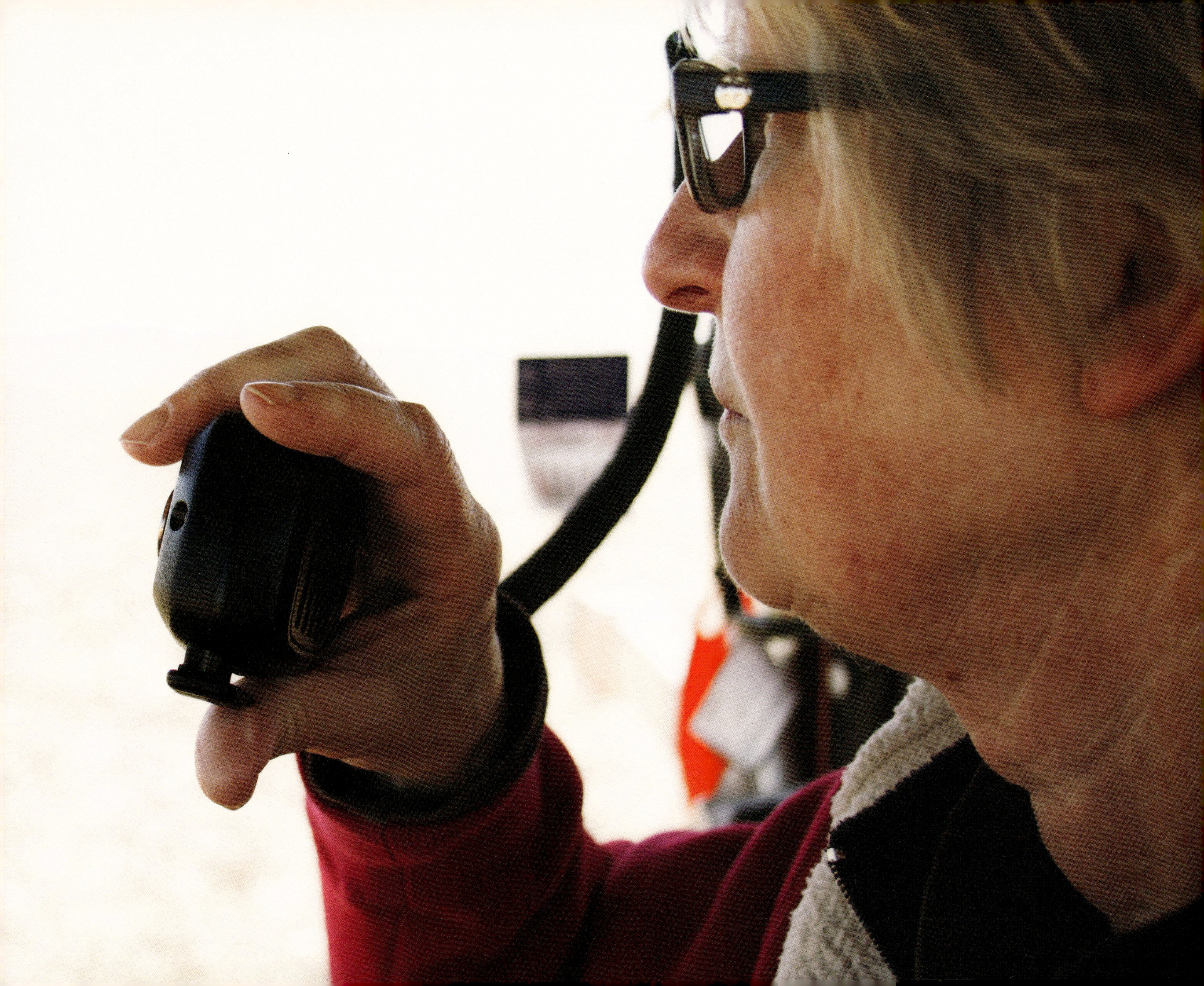

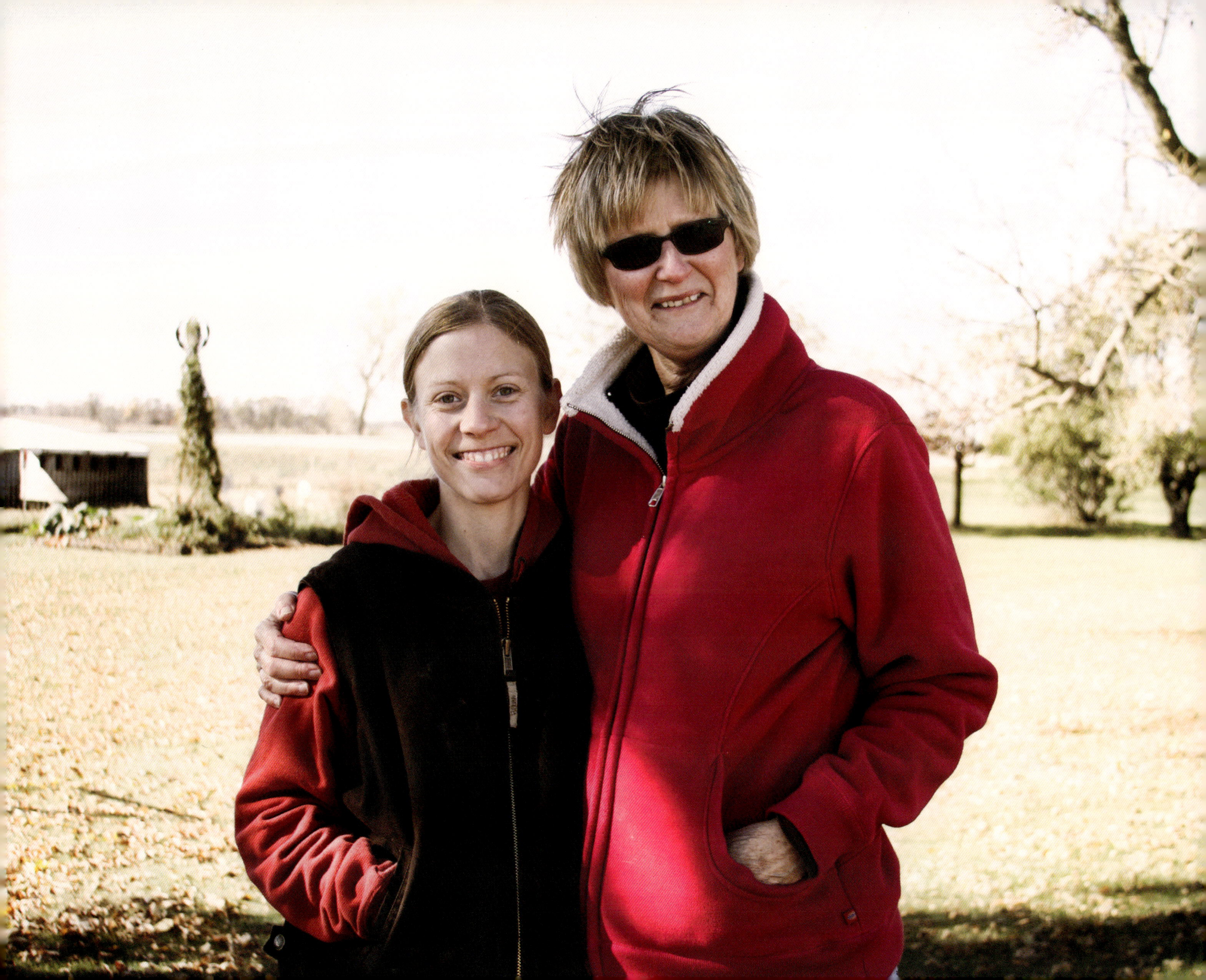

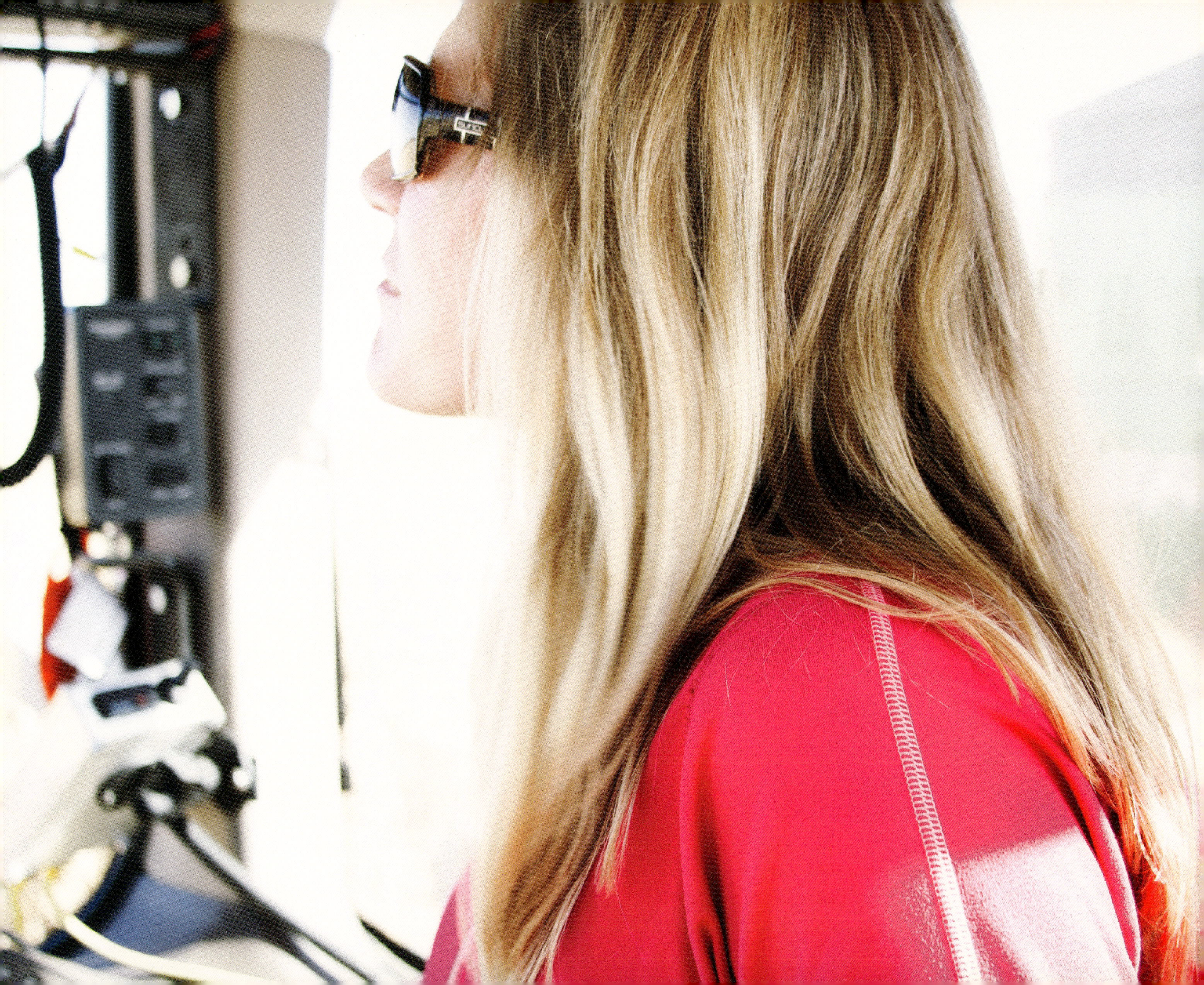

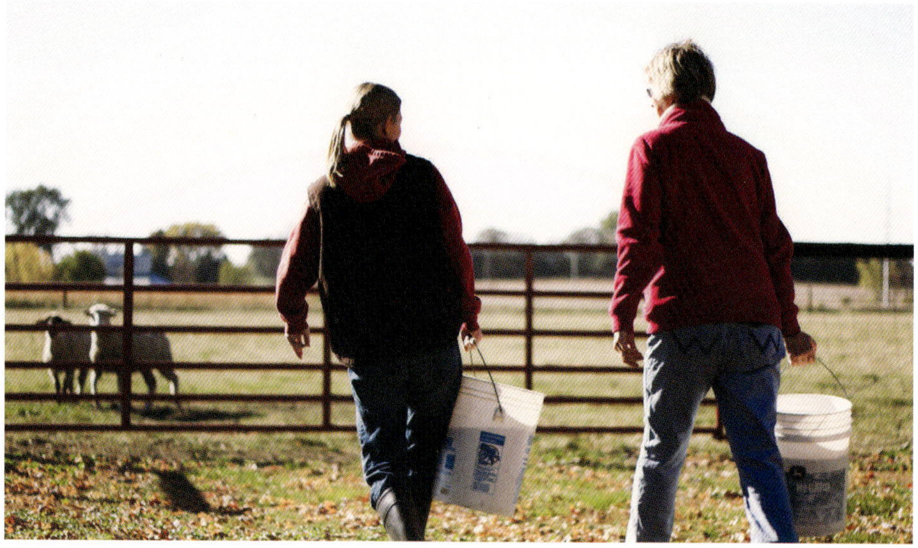
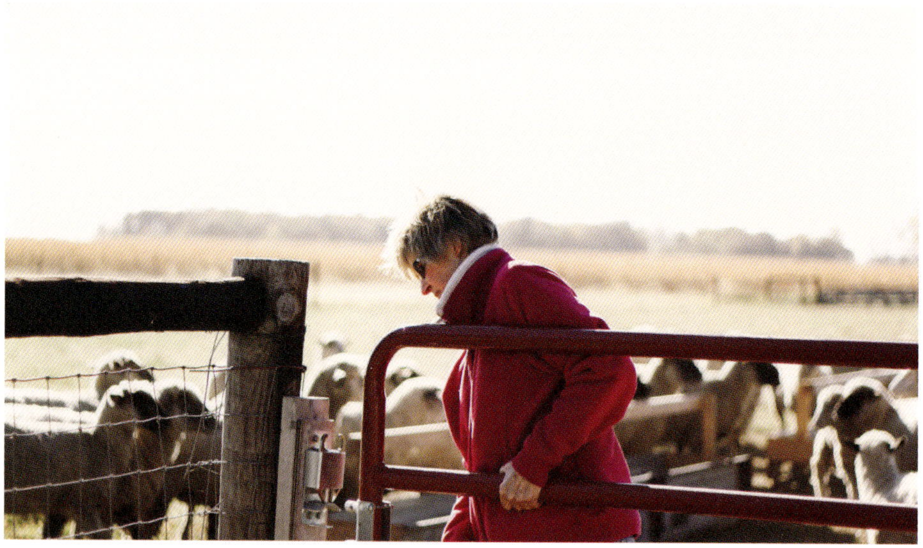
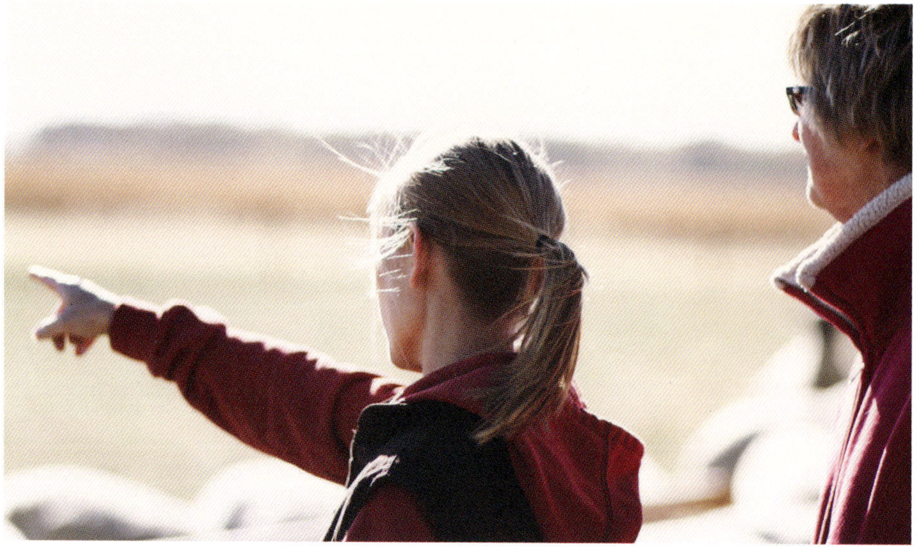
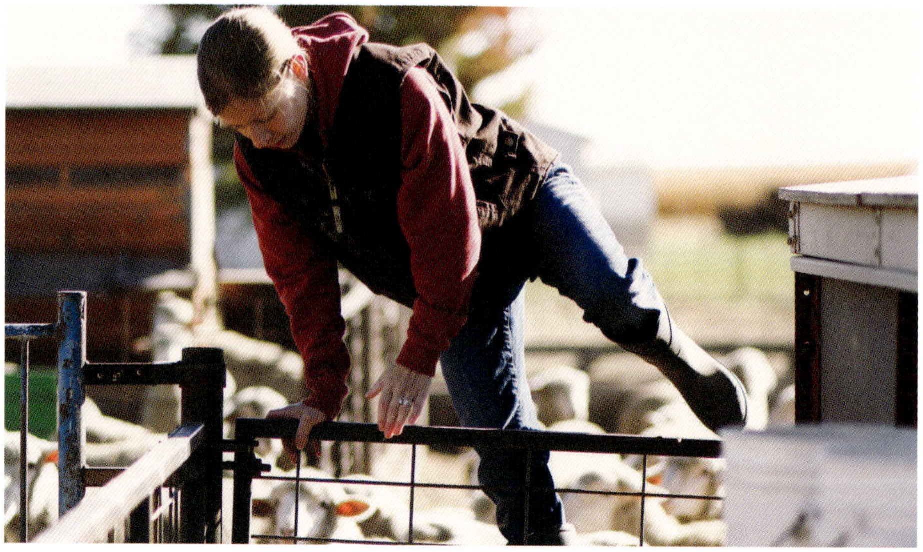

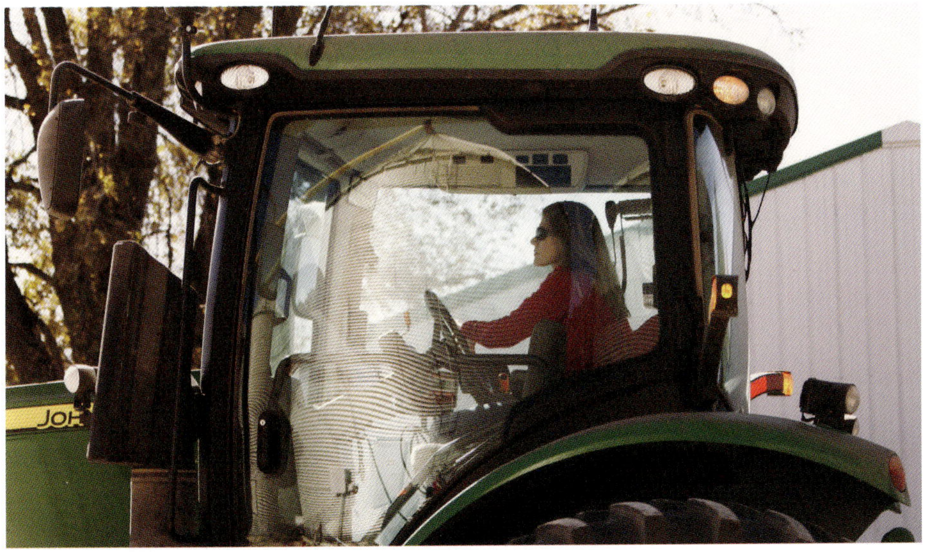
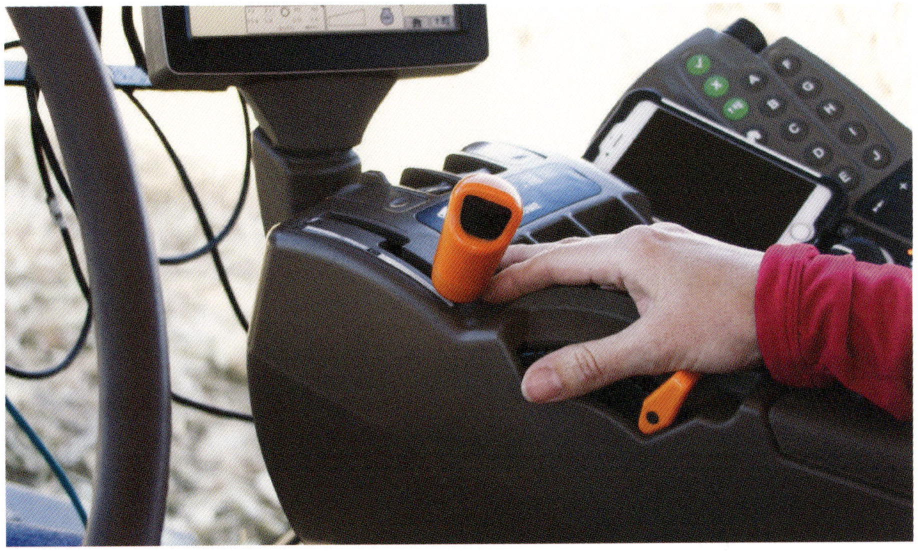
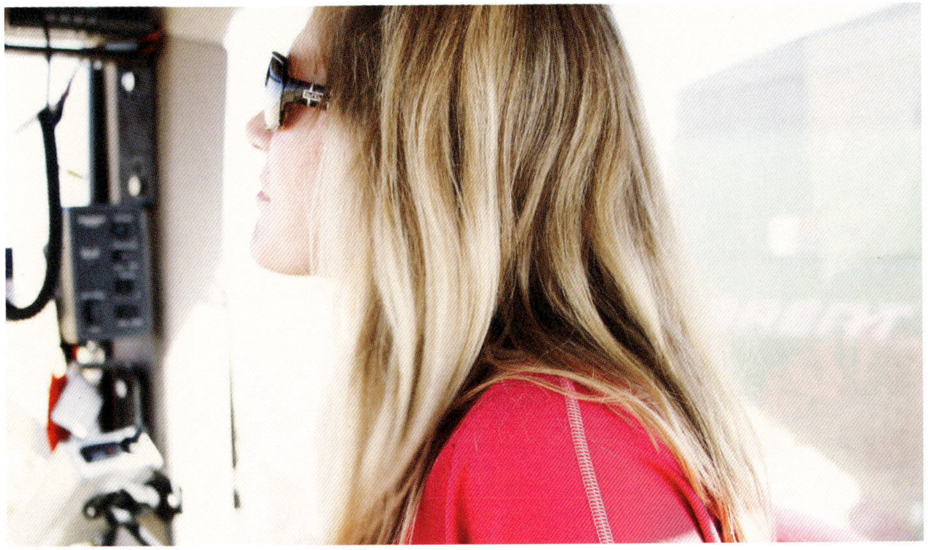
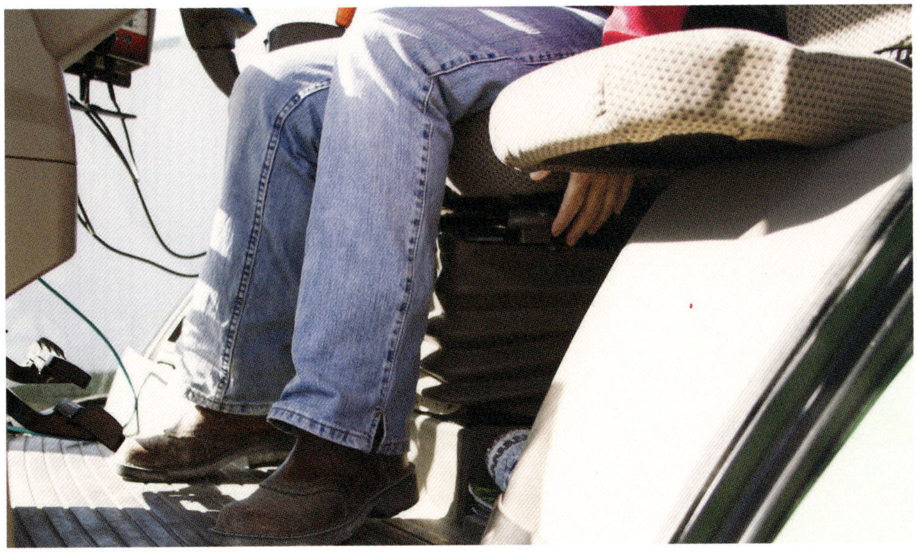

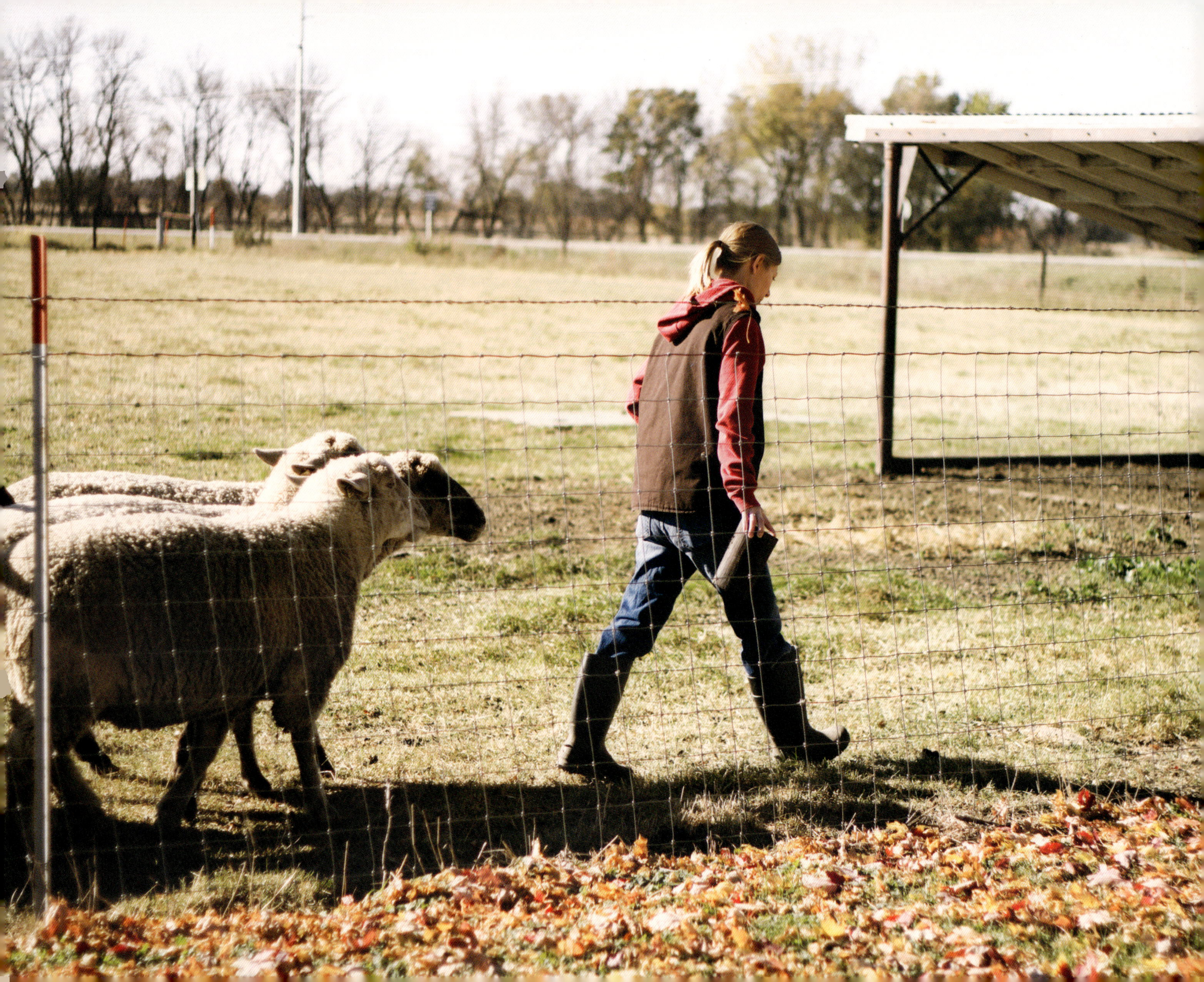

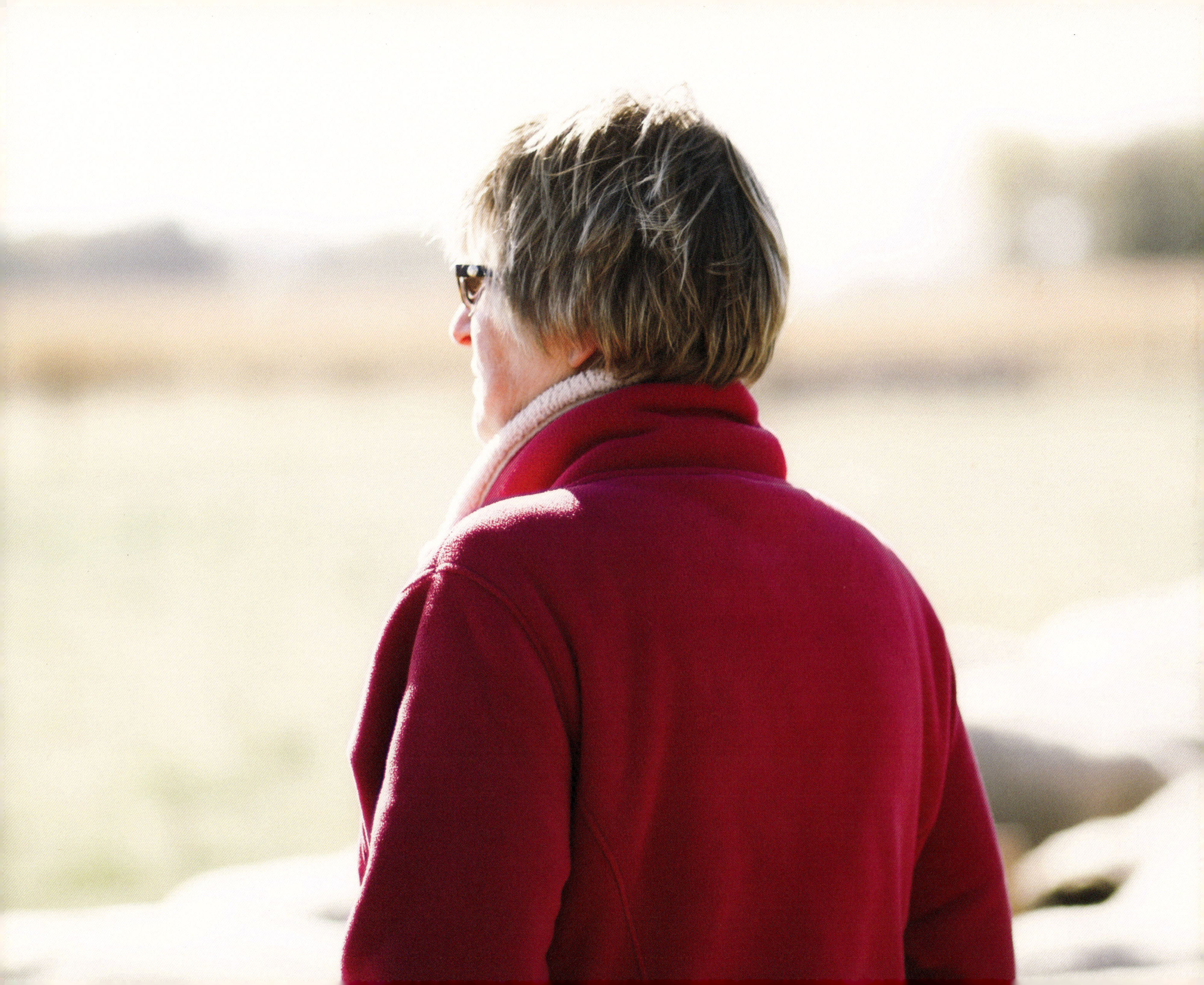

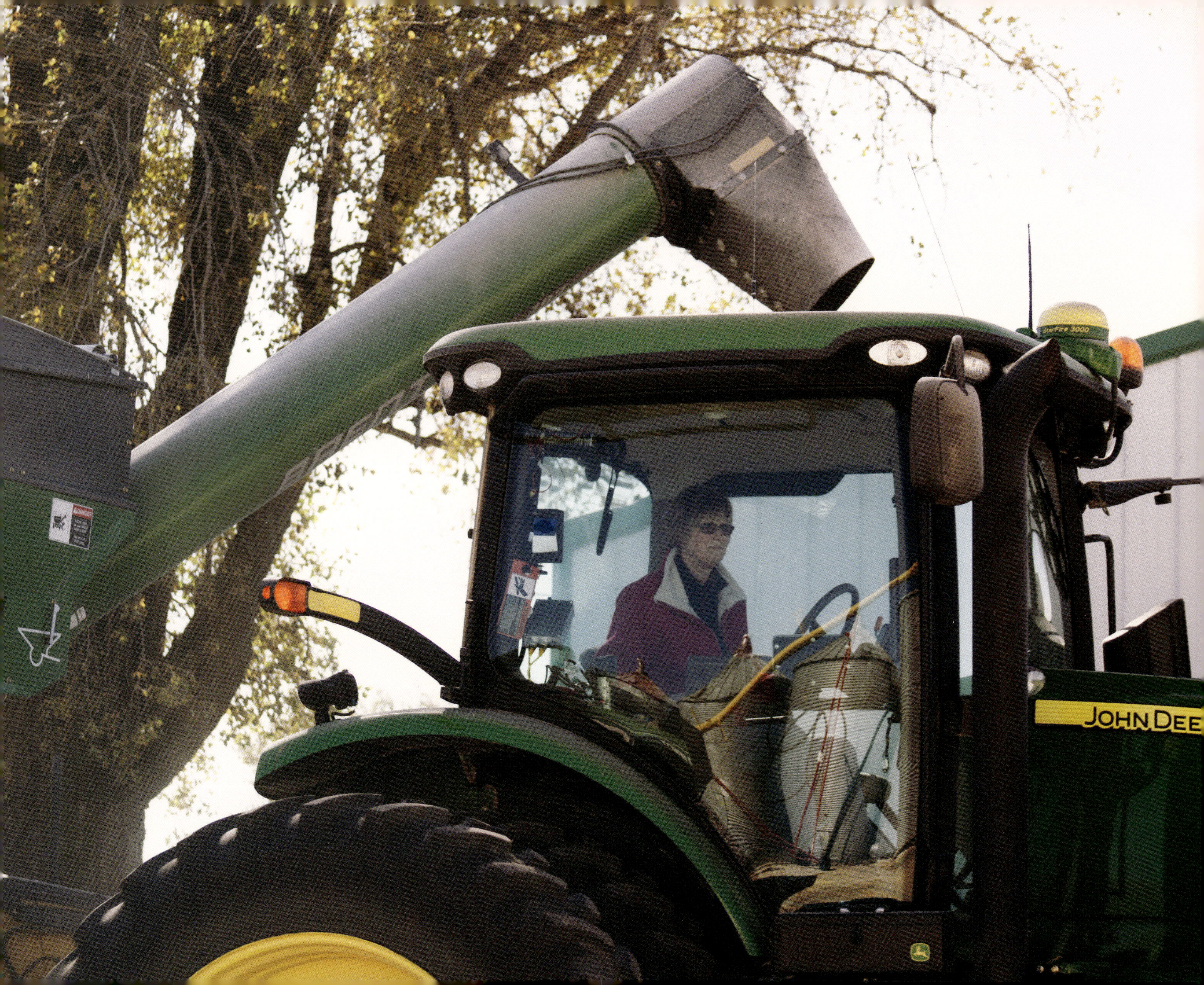
JOHN DEE

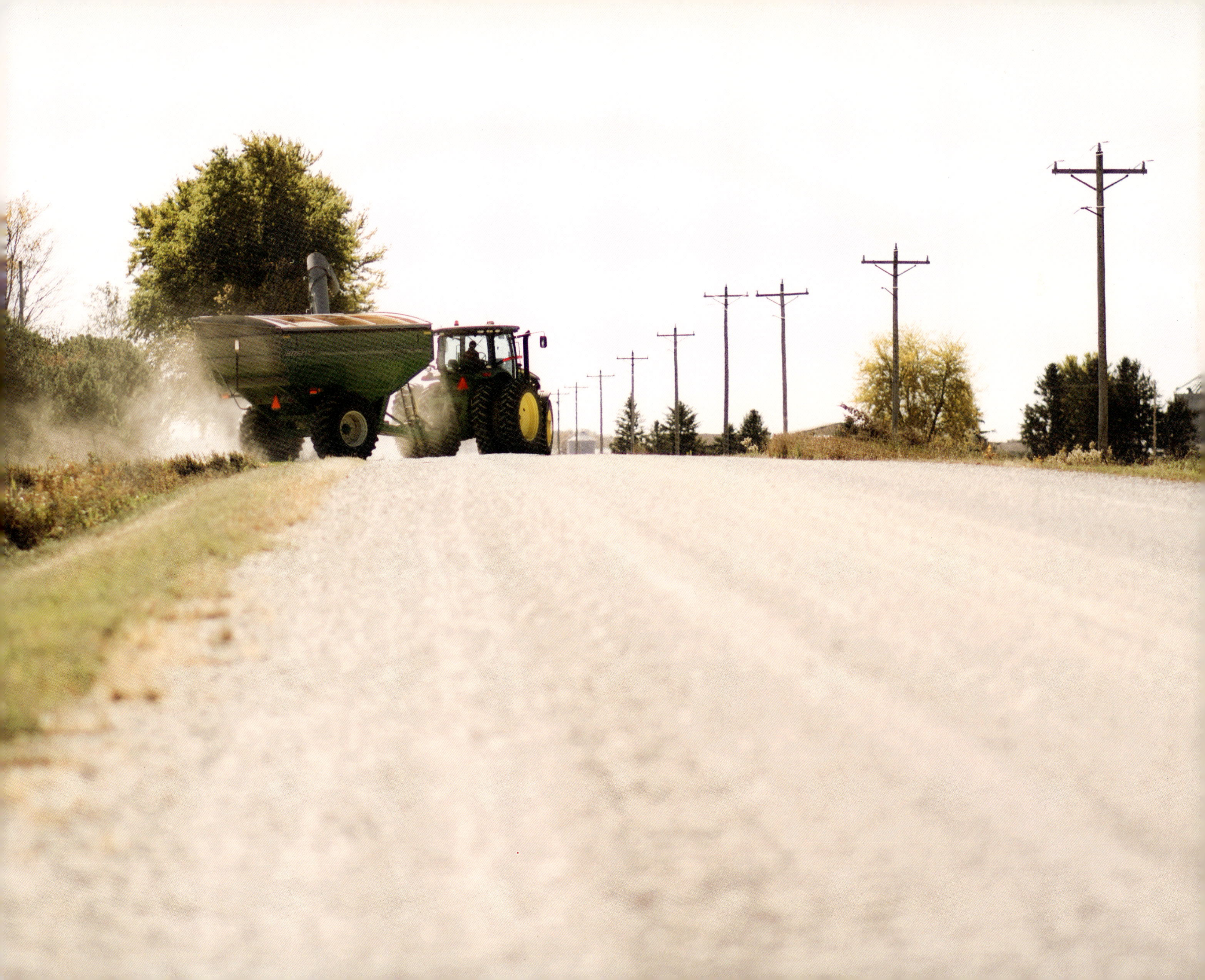

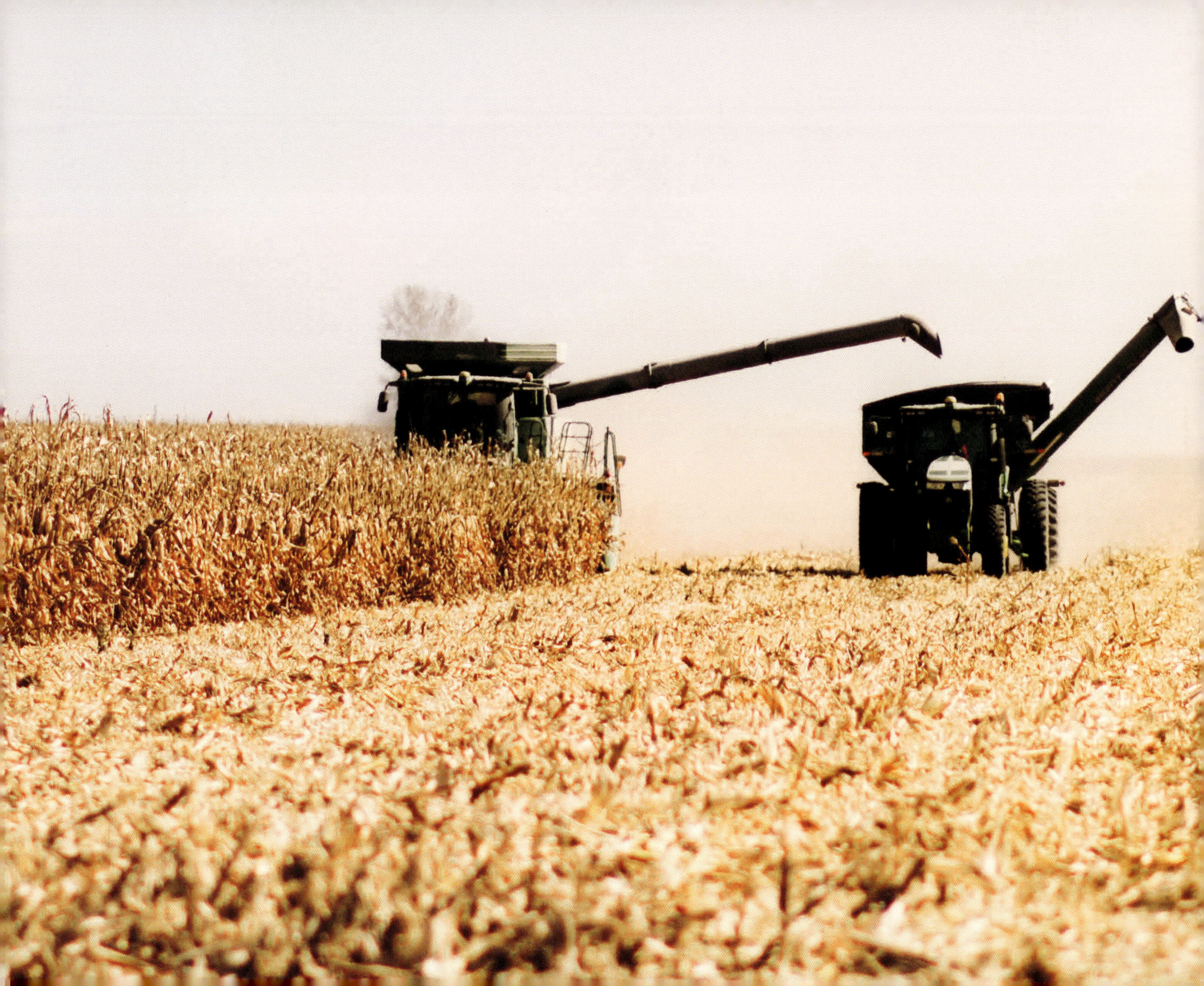

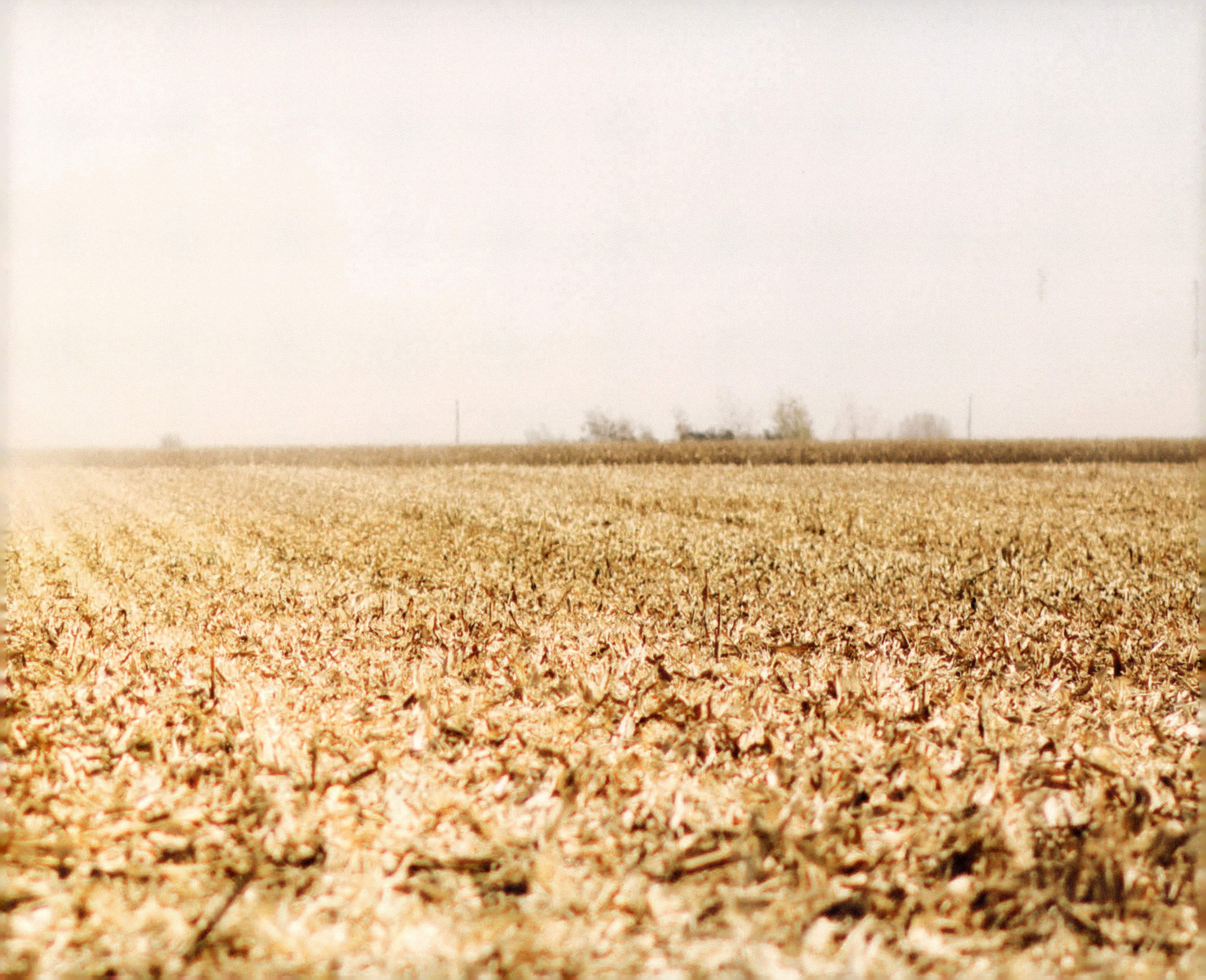

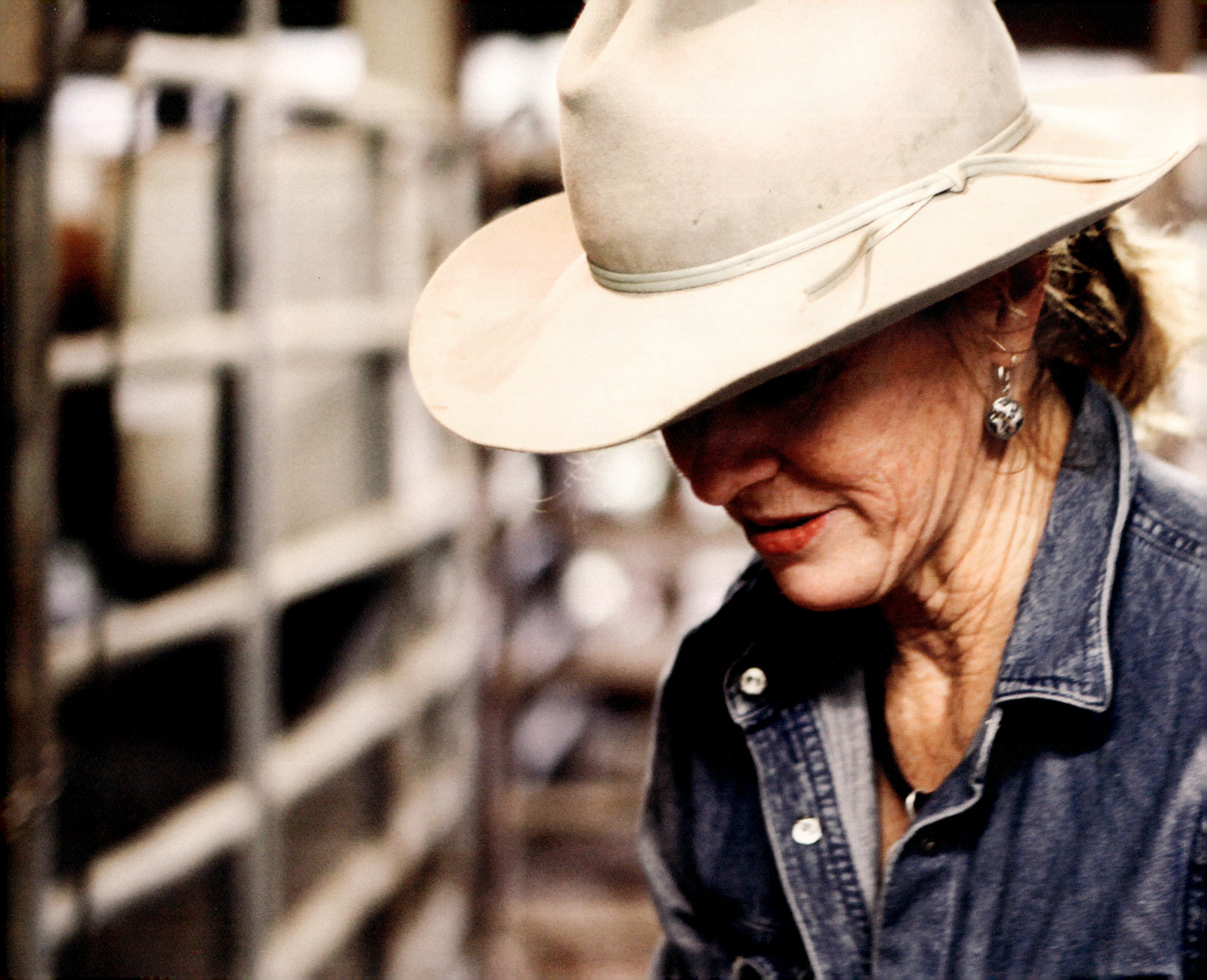

RENEÉ

STRICKLAND FAMILY RANCH • FLORIDA

Reneé Strickland is a fourth generation cattle rancher in Florida. Her father was the manager of one of the largest ranches in the state and she grew up riding and working cattle with him.

Reneé also loves horses and uses them to work the cattle and to play polo, one of her passions.

One of her other passions, traveling, inspired her to become involved in the exporting business. She is one of the few women who are in the export business and serves as president of the Livestock Exporters Association of the U.S.

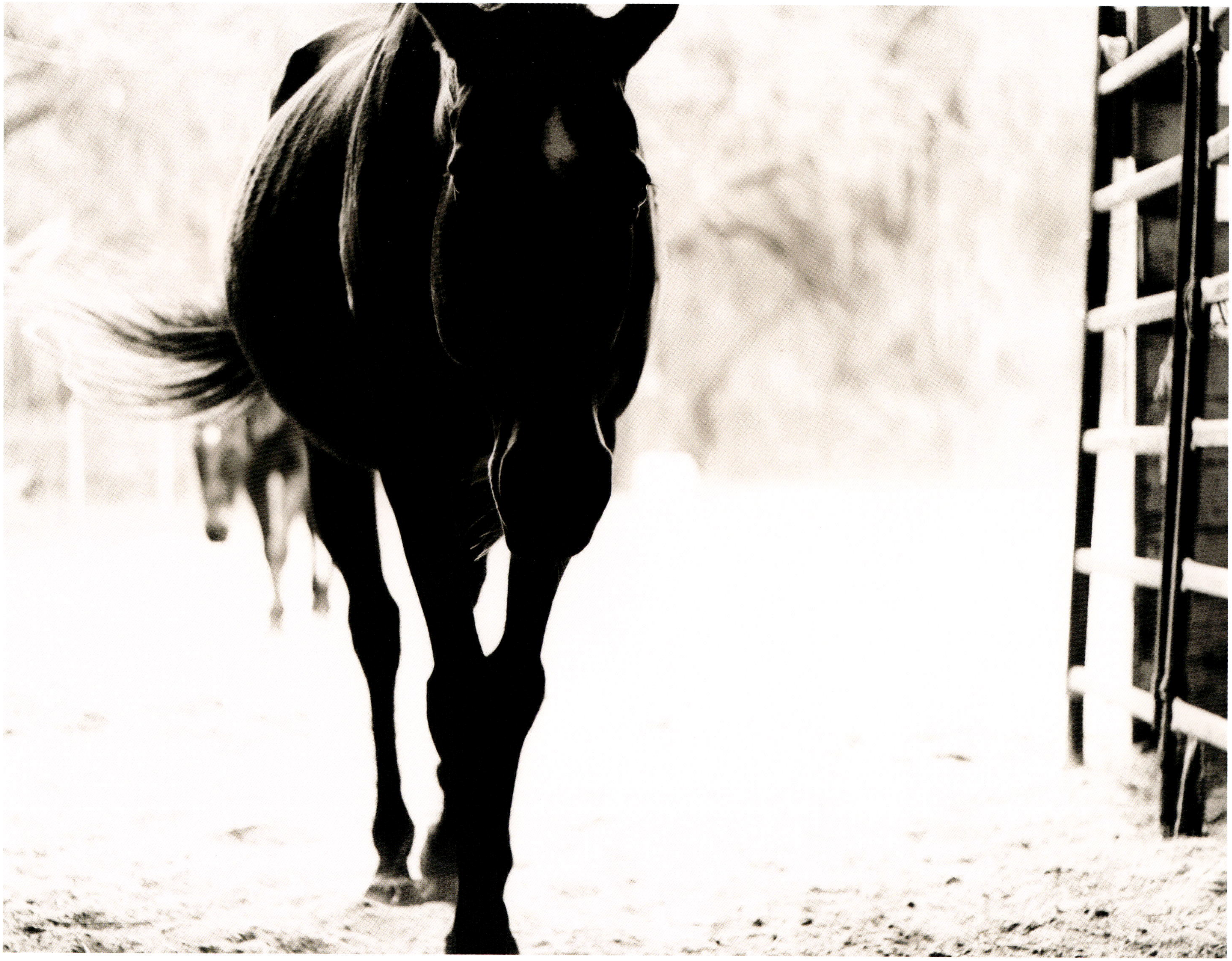

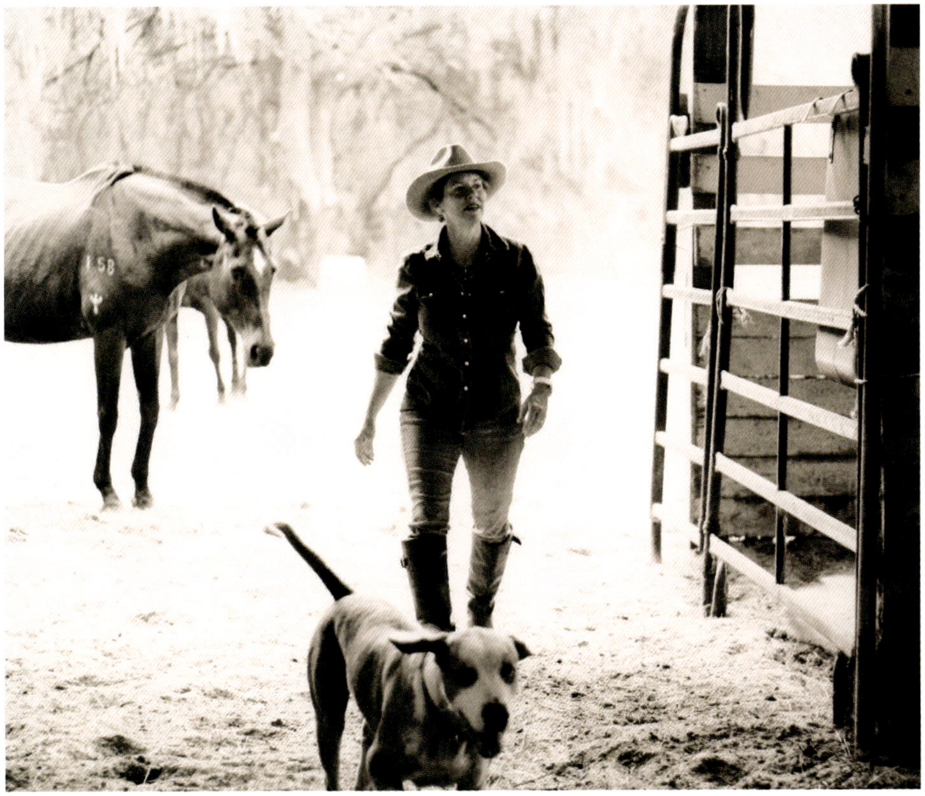

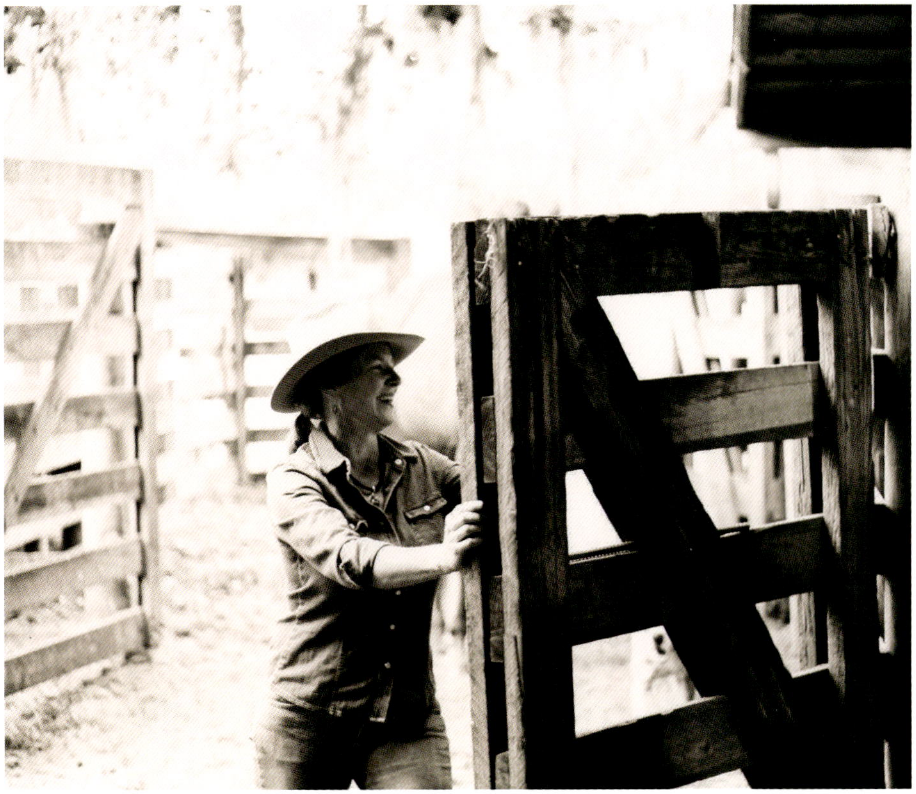

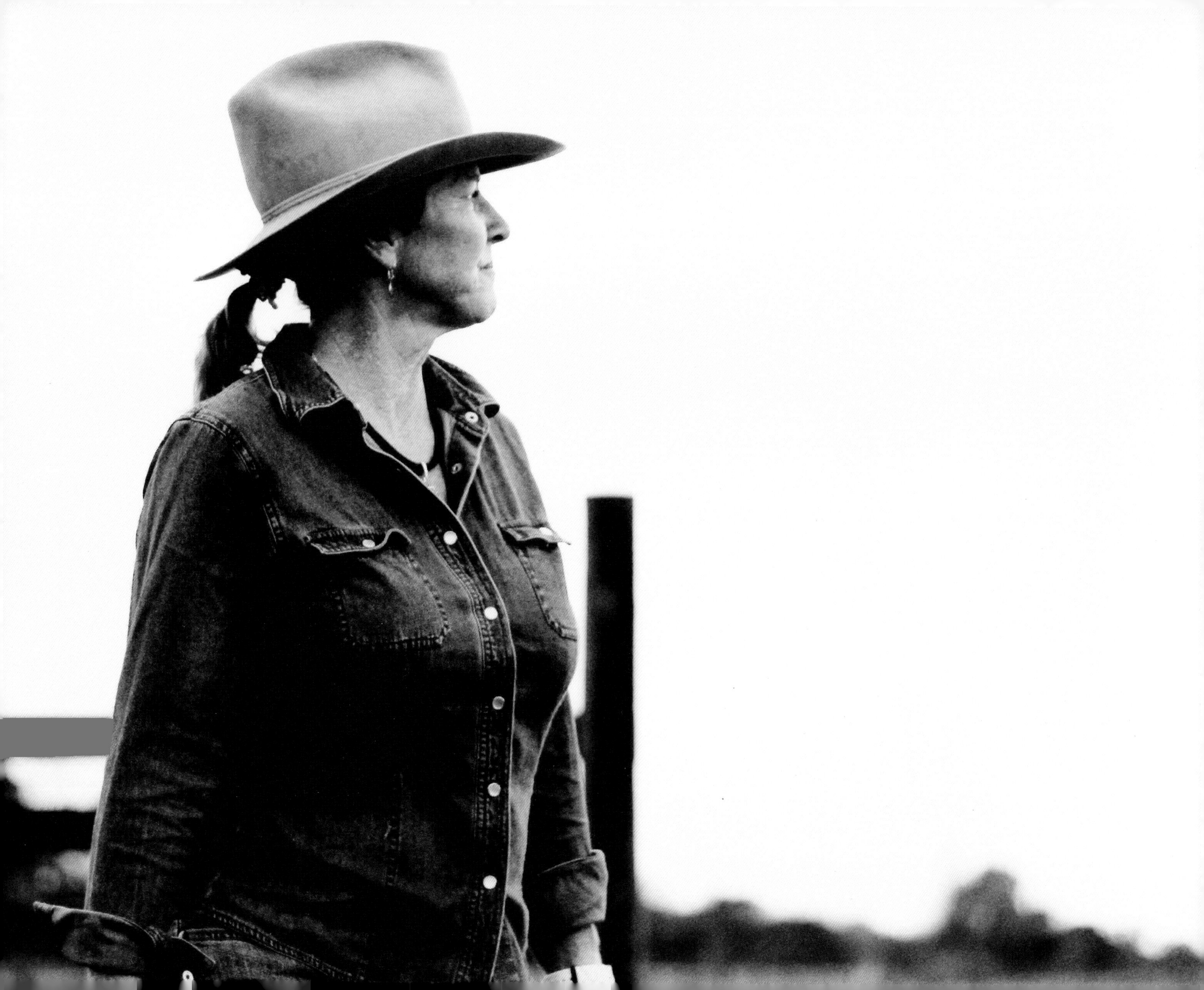

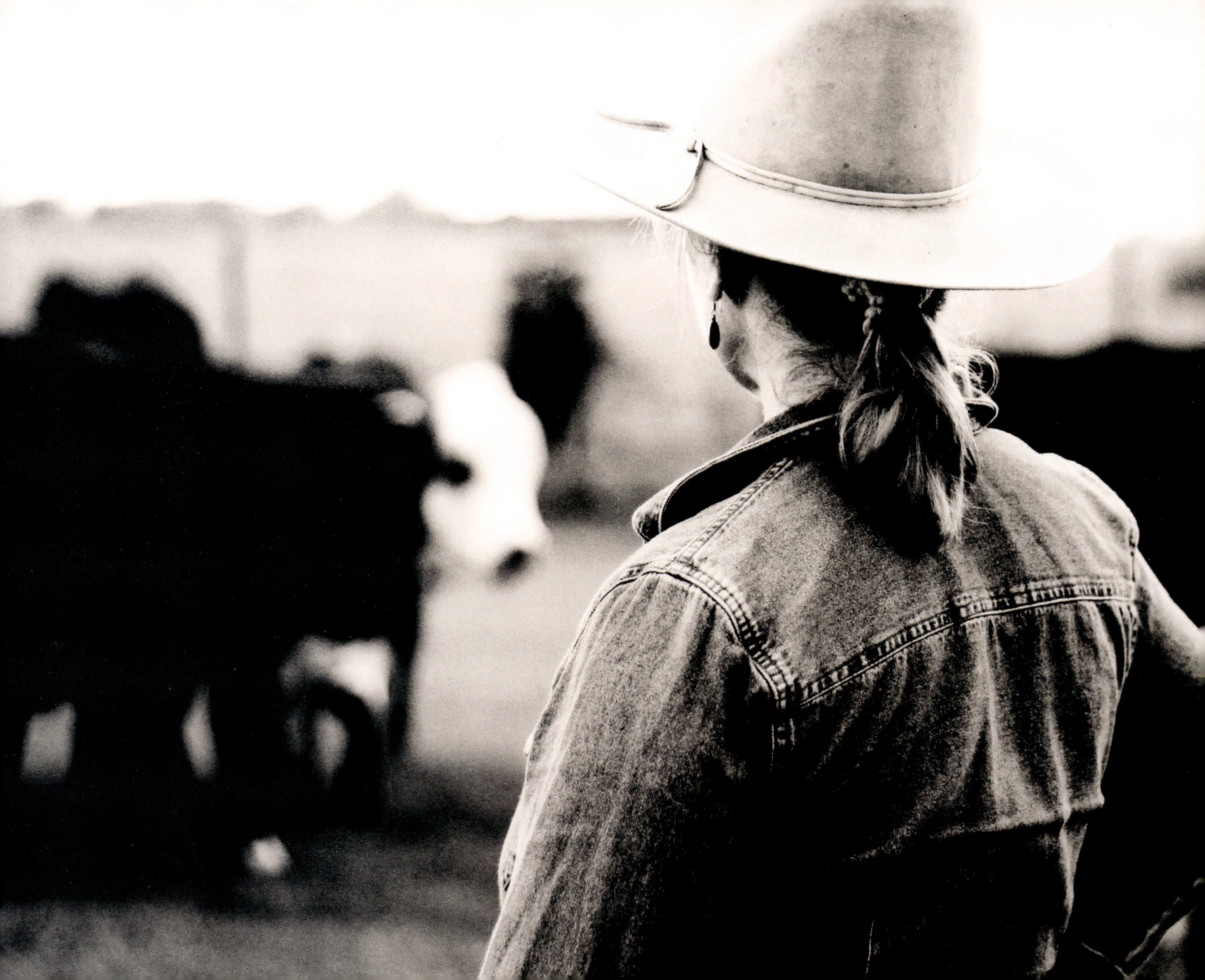

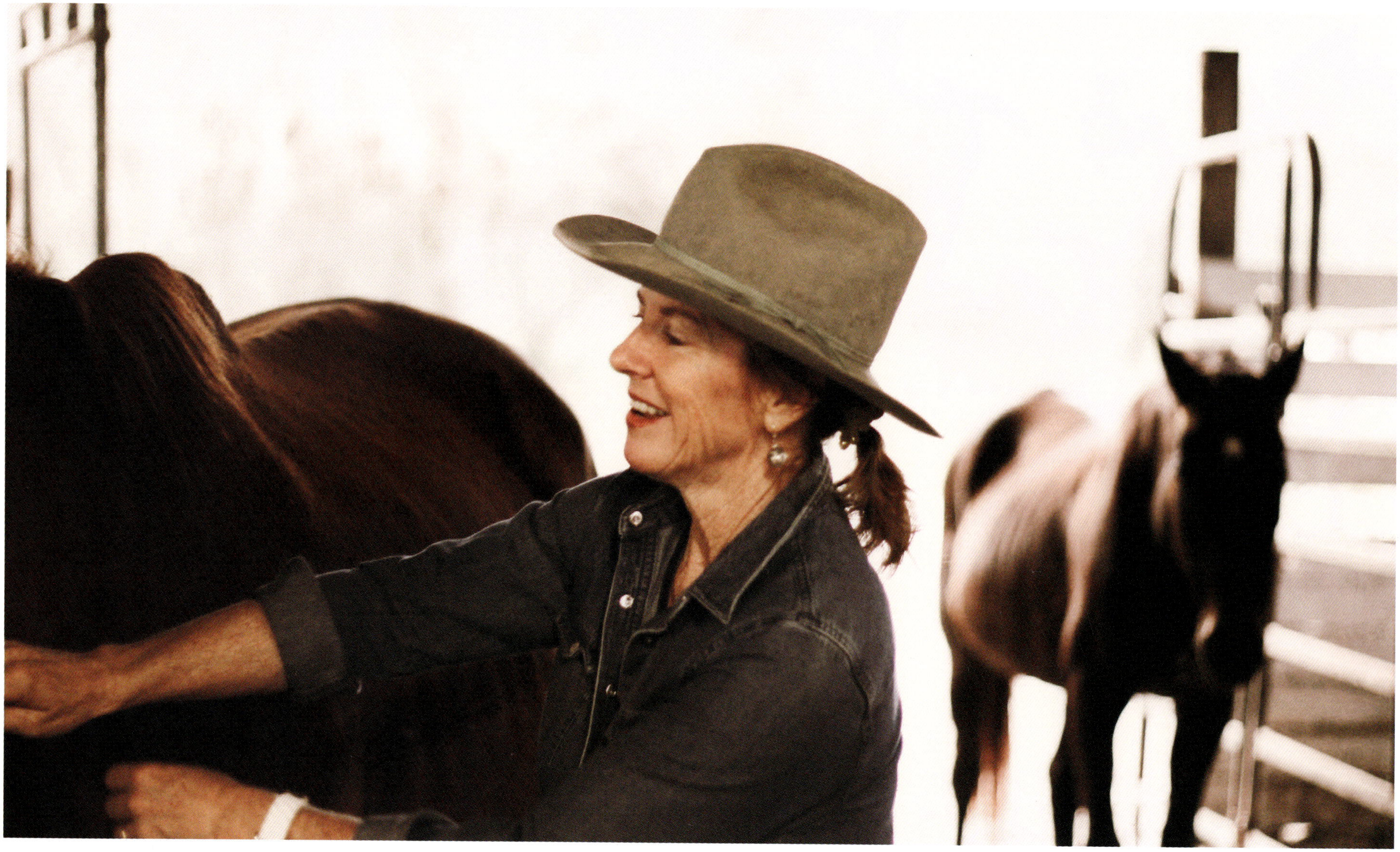

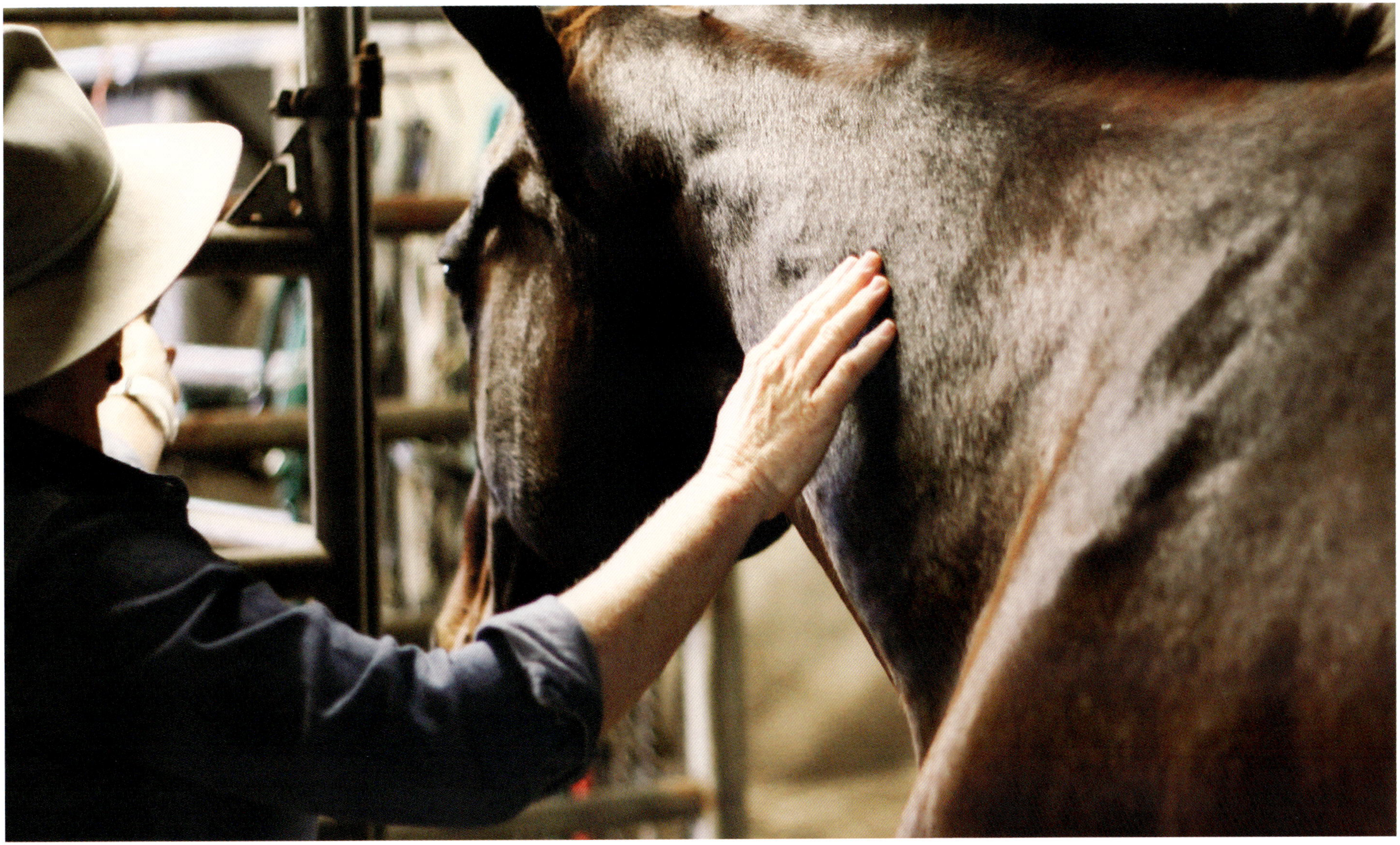

SARA

HOLLENBECK RANCH • MONTANA

Sara Hollenbeck is a sheep rancher with her husband and his family near Molt, Montana, 35 miles northwest of Billings.

Sara was always interested in agriculture growing up in California and being involved in FFA, but never imagined she'd be raising sheep in remote Montana.

To get more people interested in eating lamb and to promote Hollenbeck Ranch, Sara created High Five Meats to offer locally raised and sourced meat for consumers in Montana.

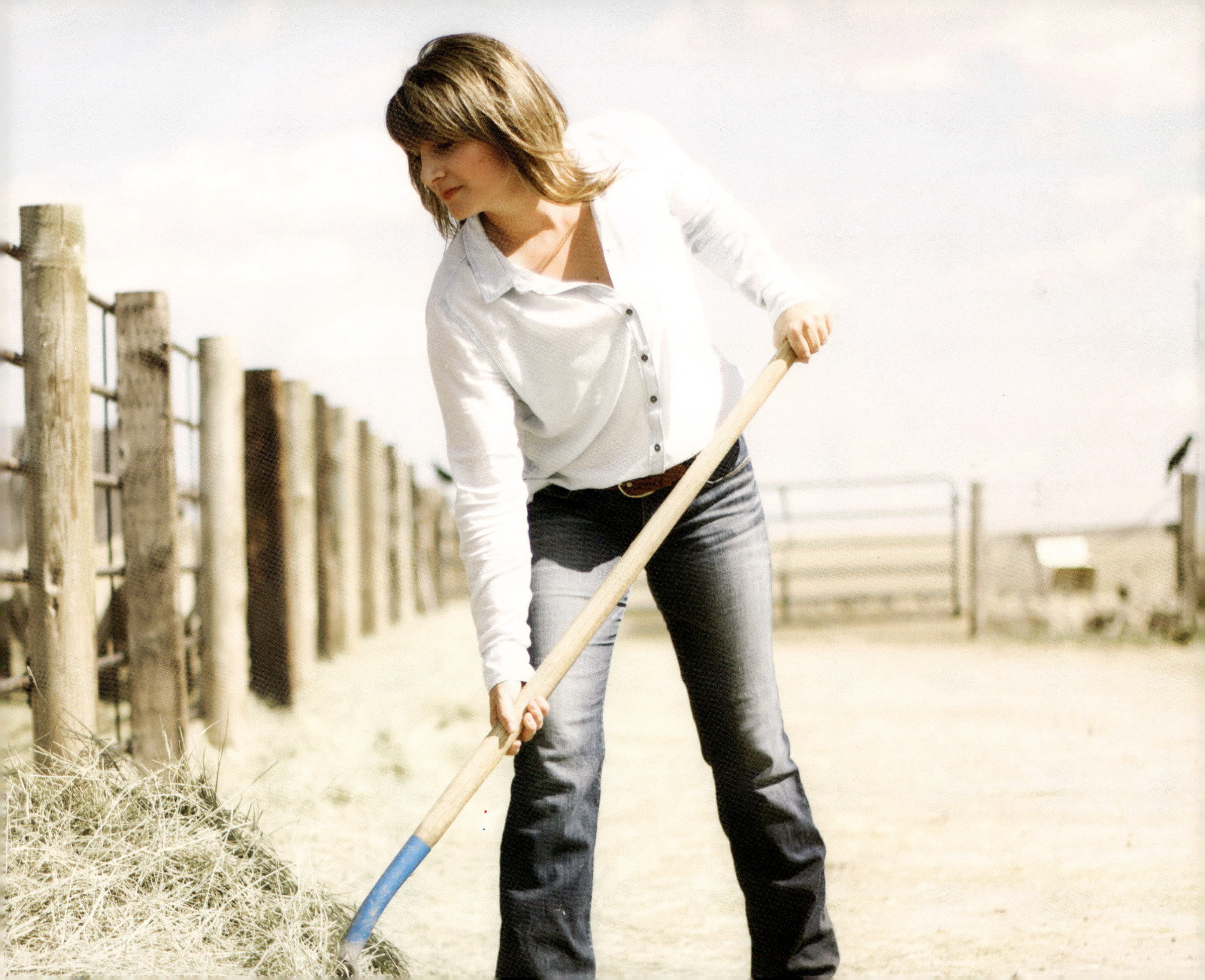

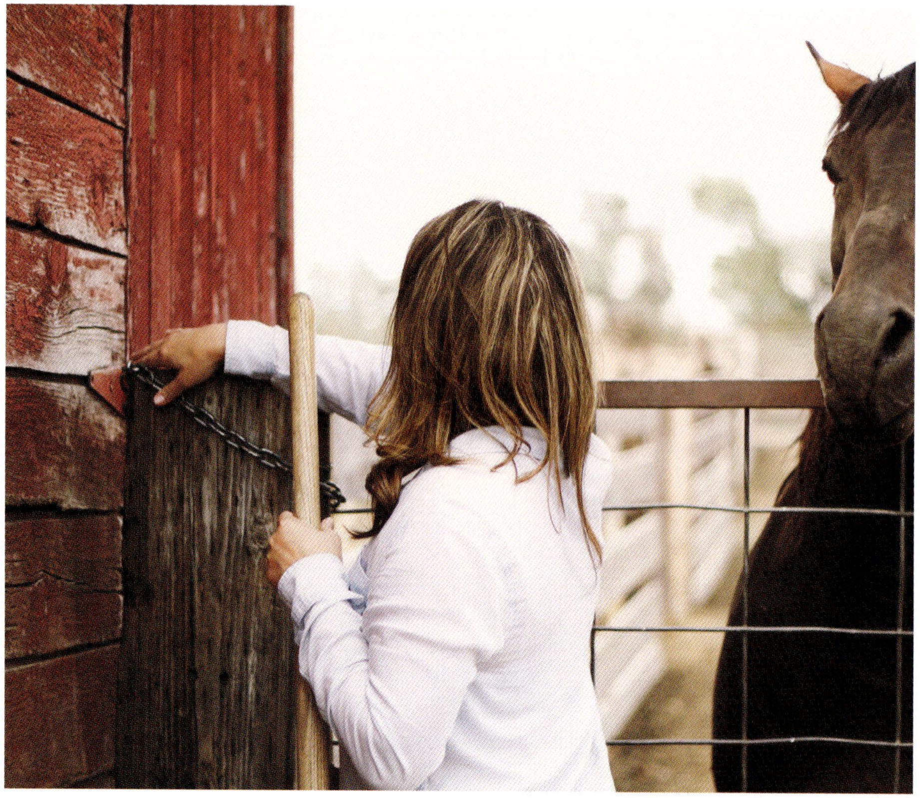
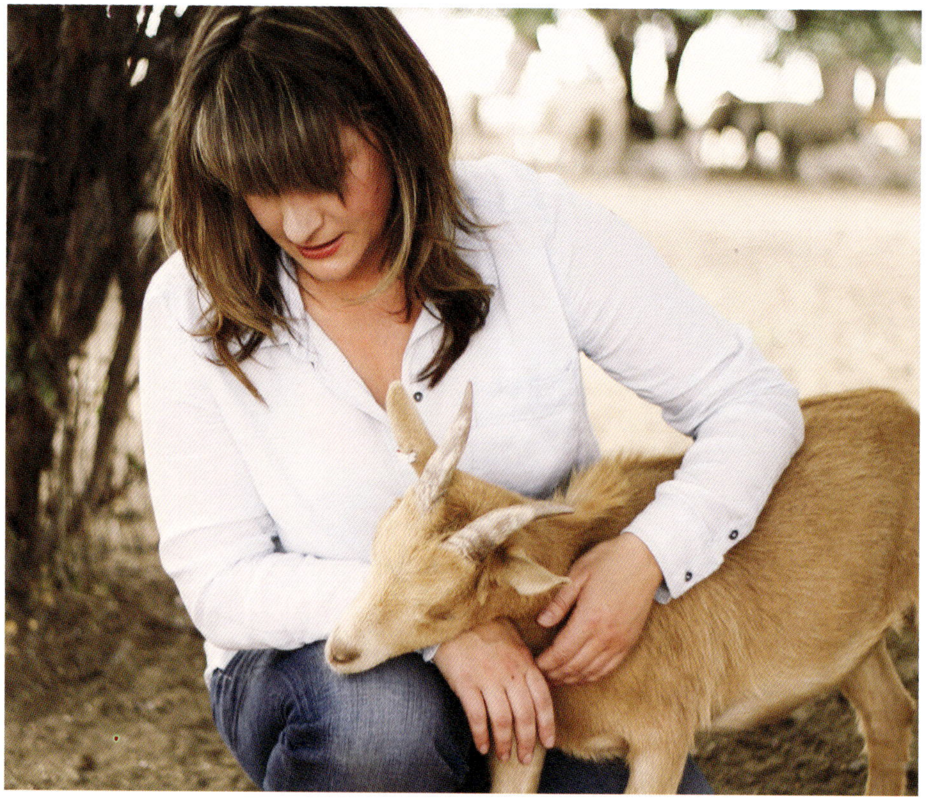

87

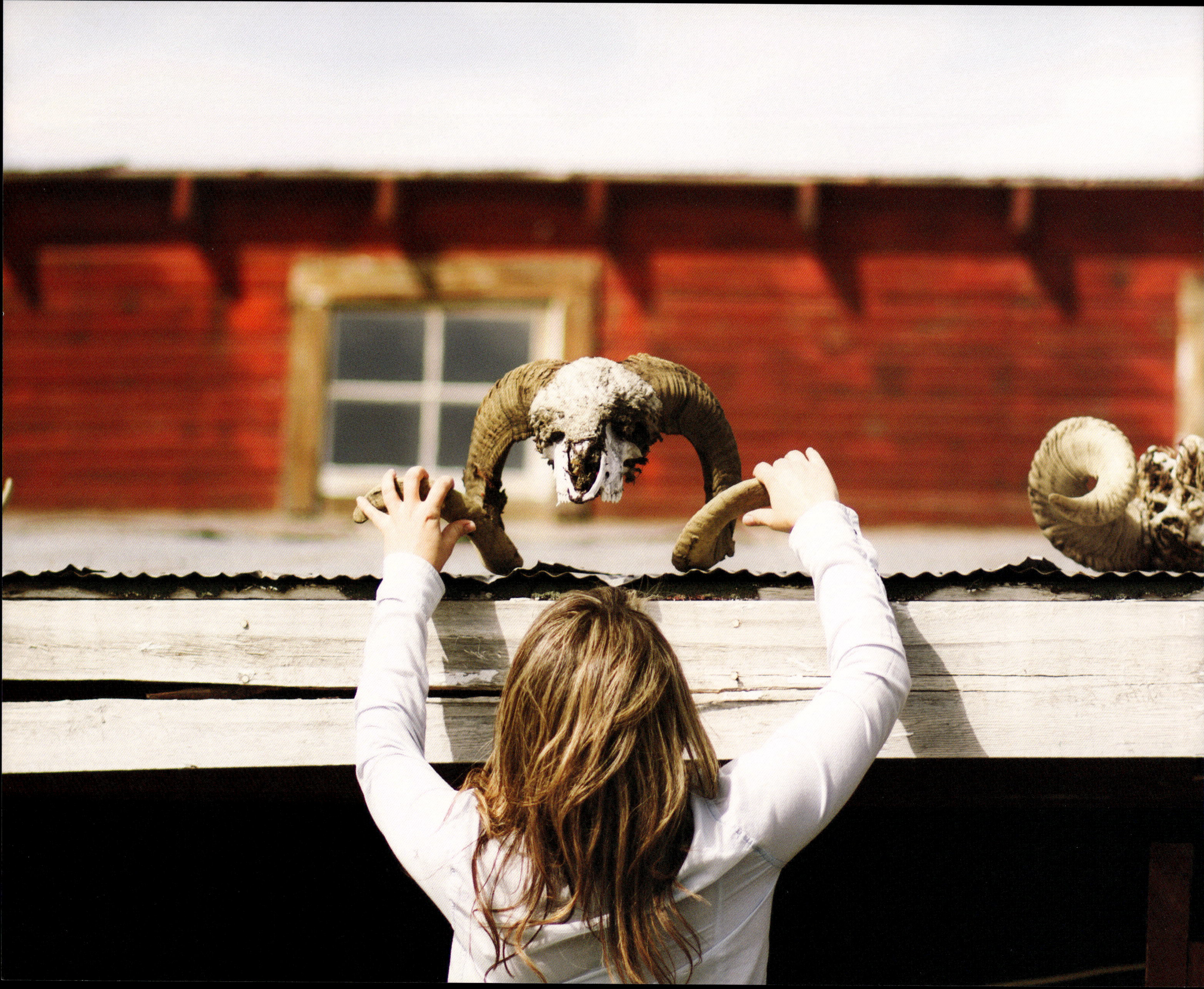

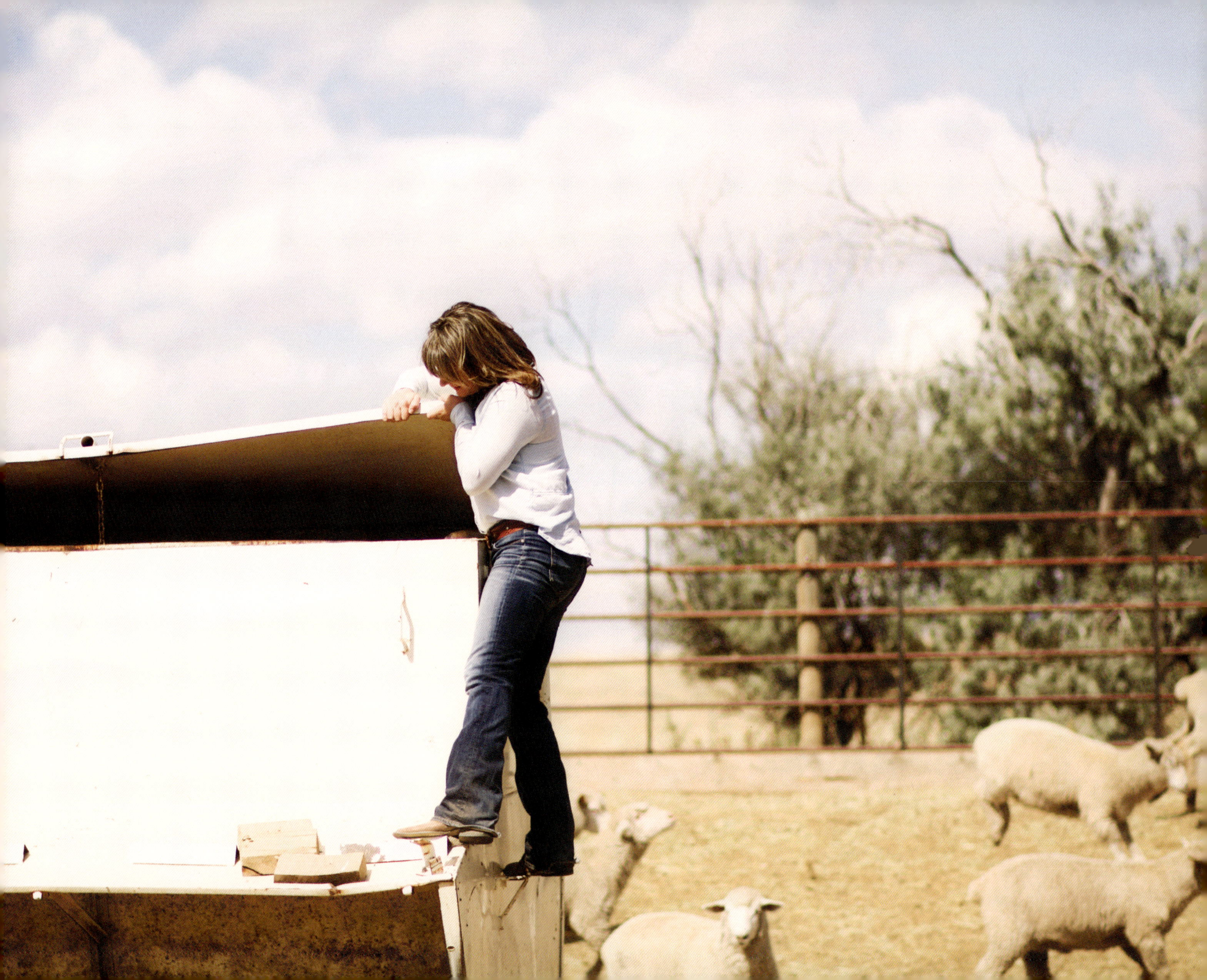

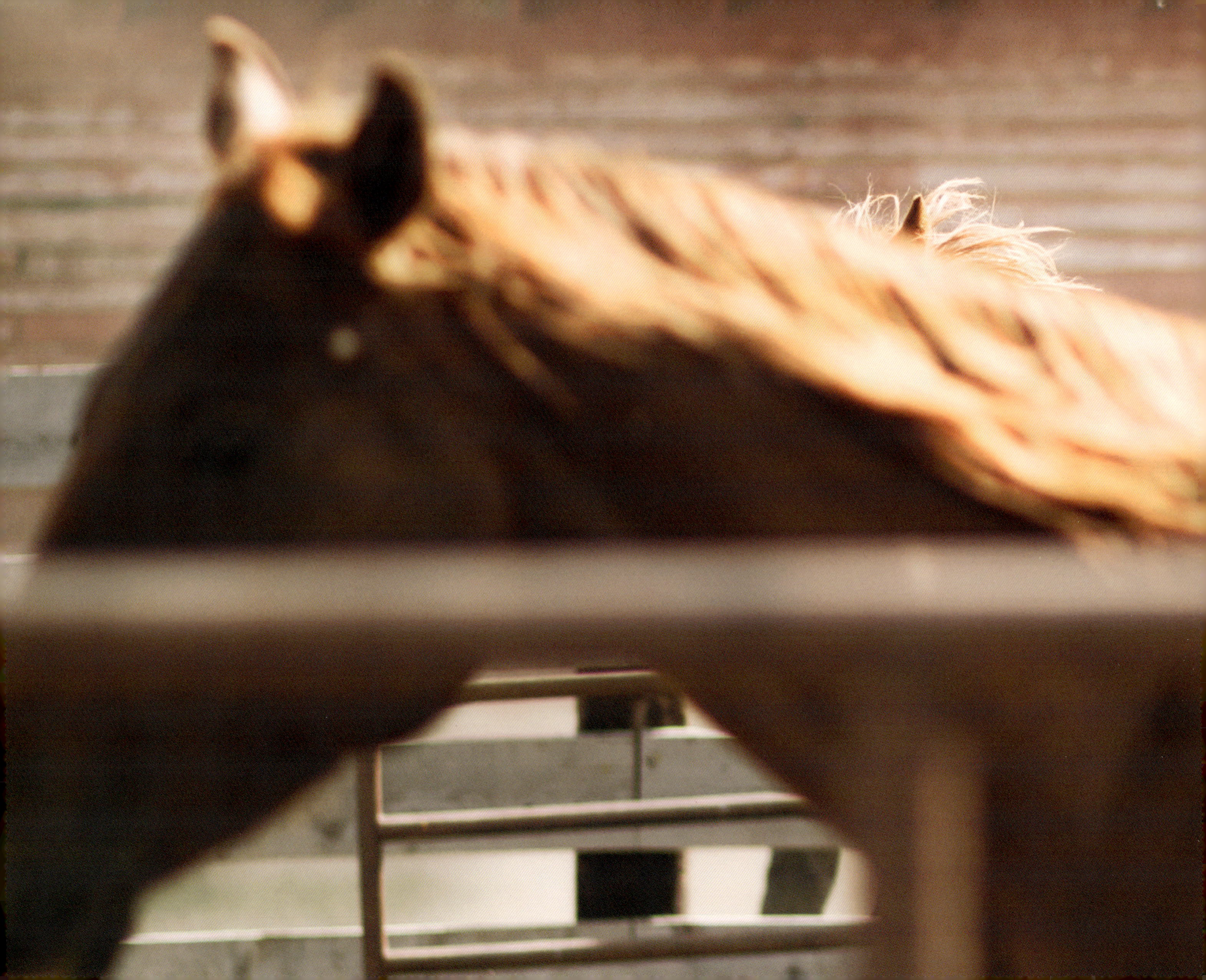

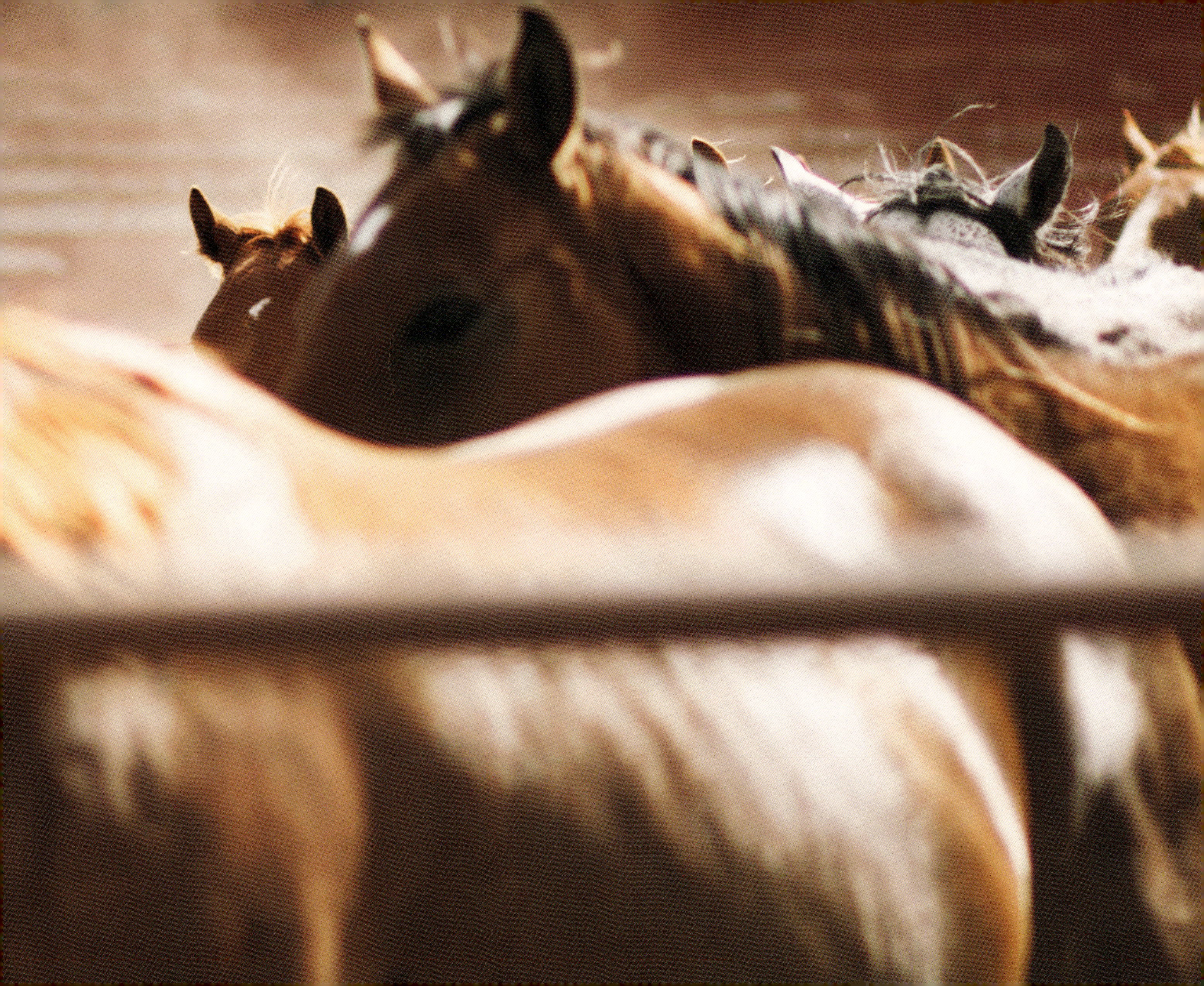

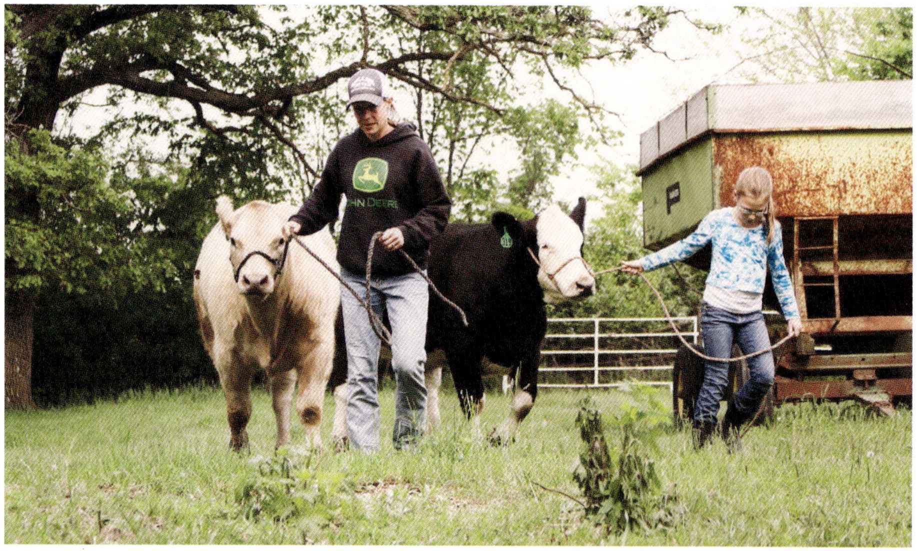

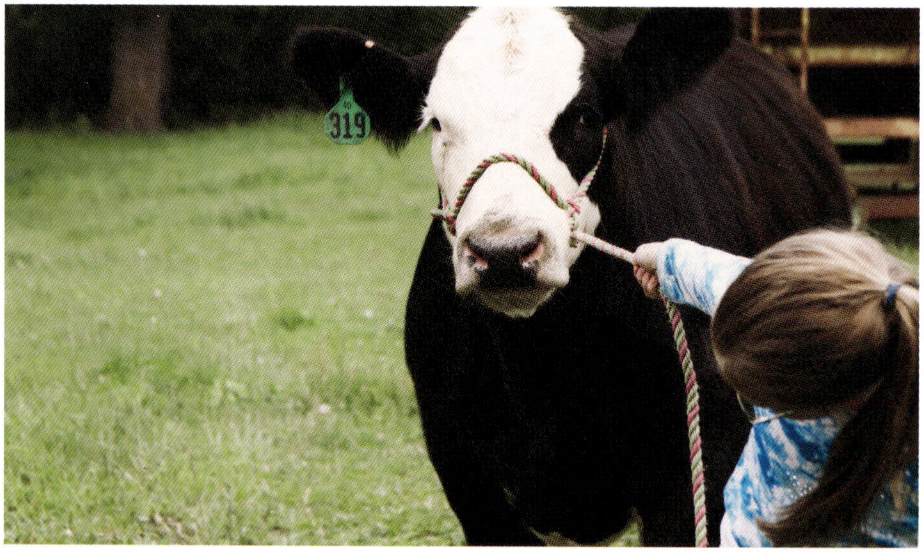
319

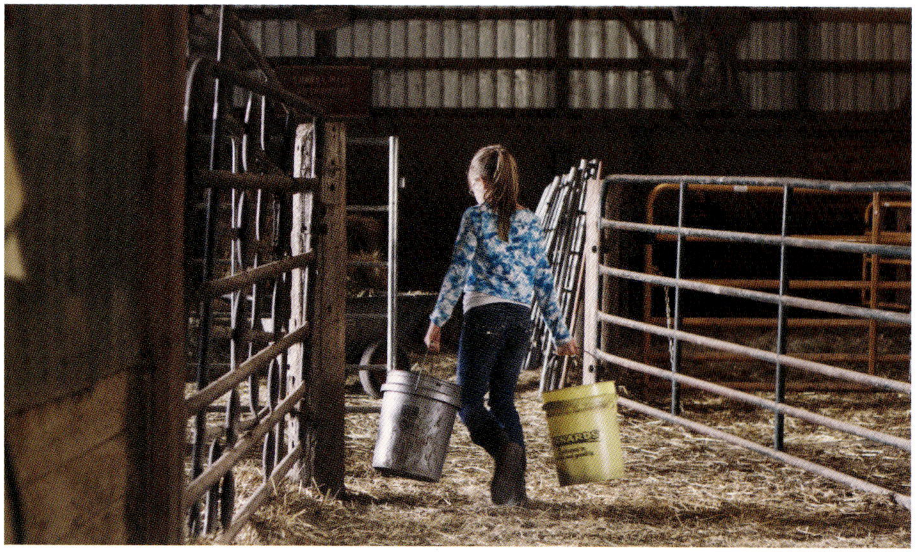

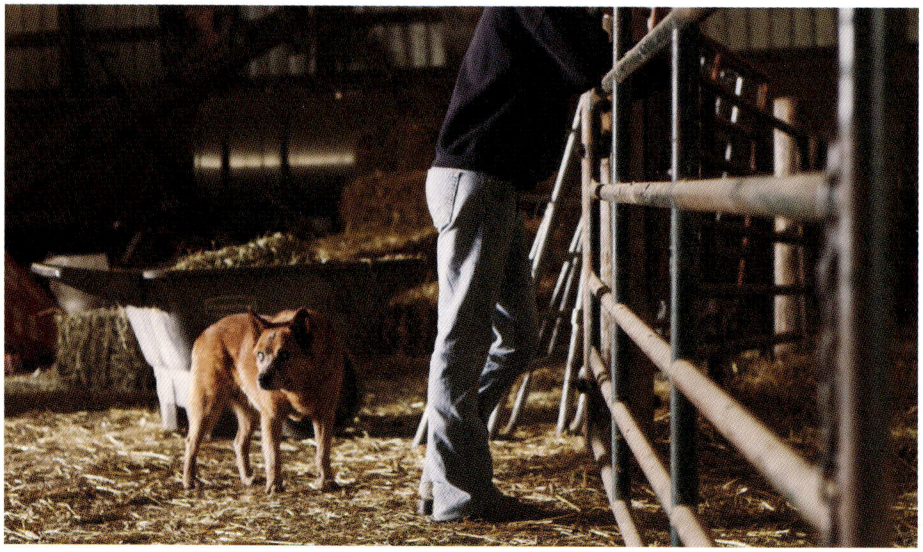

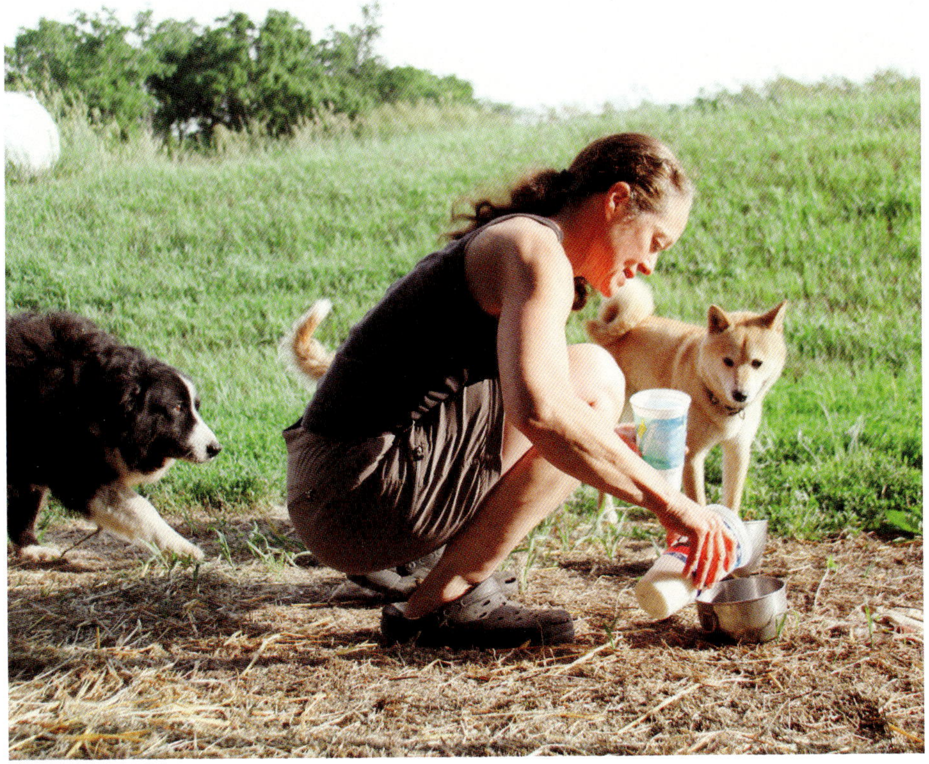

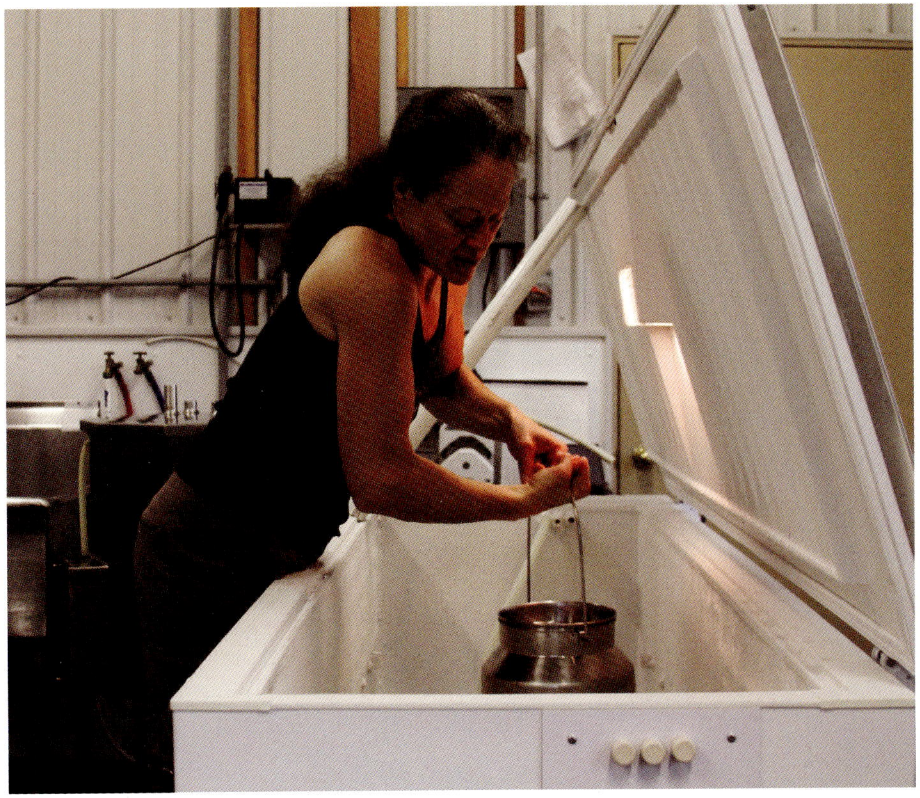

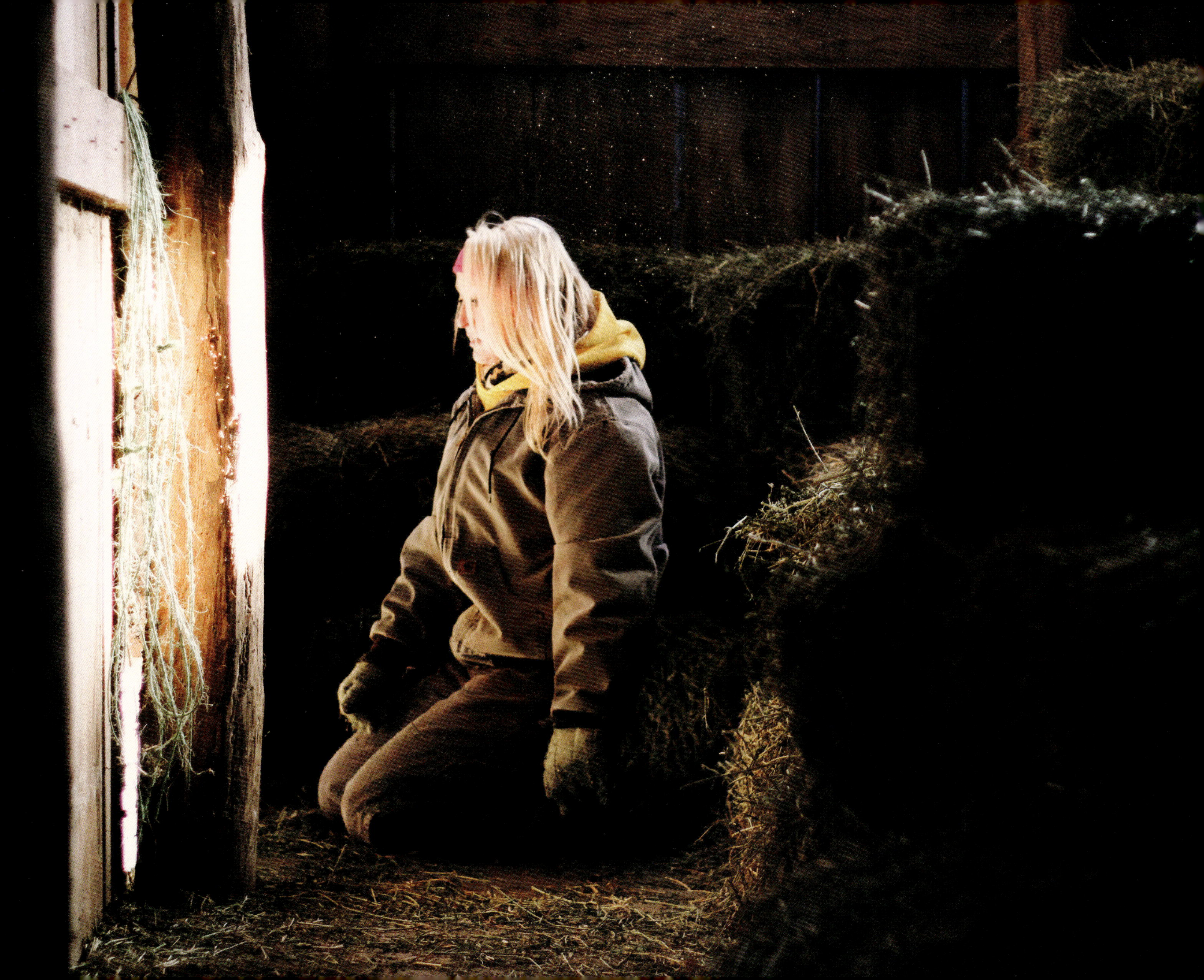

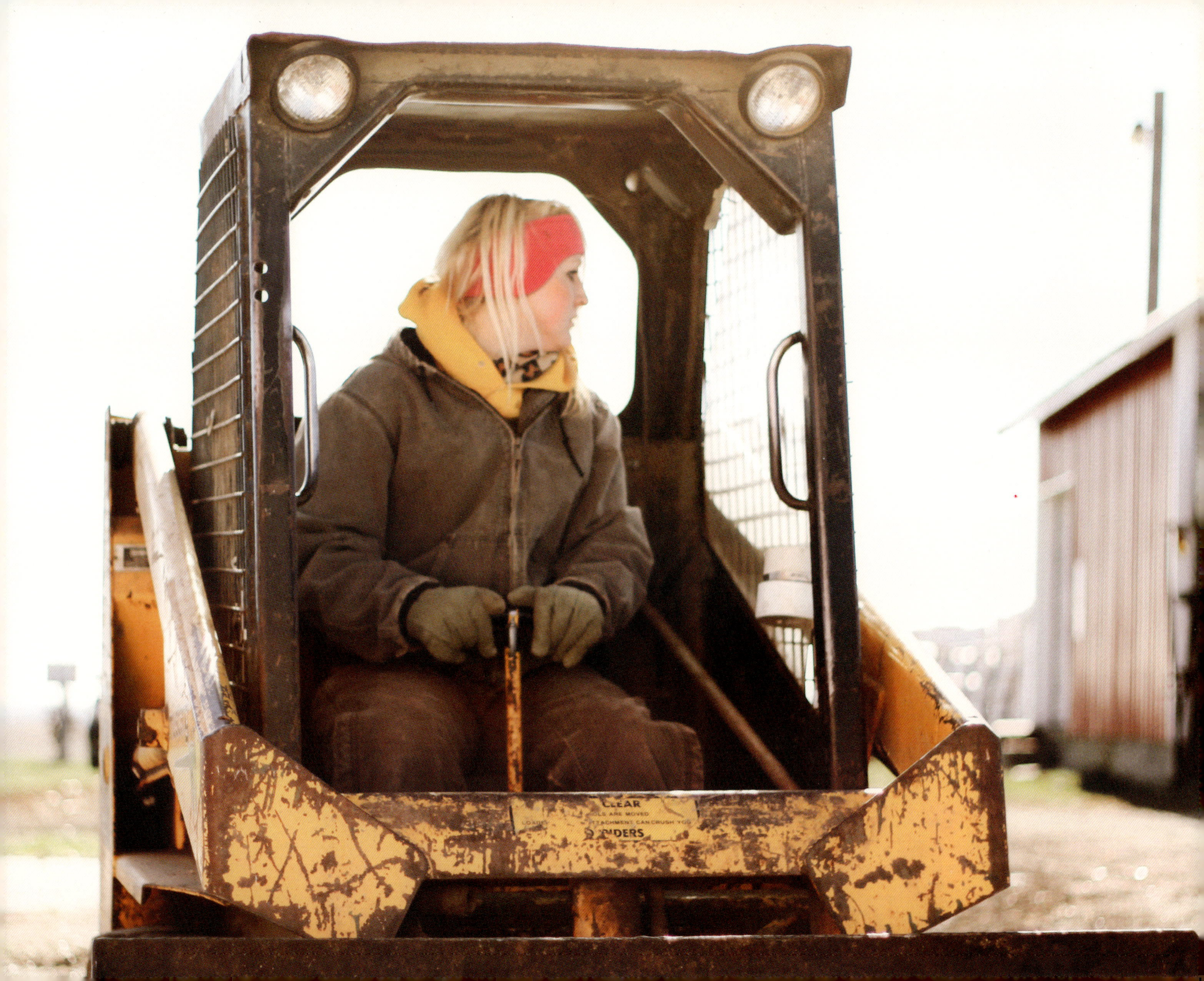
CLEAR

FarmHer
THE DOCUMENTARY

In 2013, Marji Guyler-Alaniz saw a commercial about American farmers. While the commercial moved her, she felt something was missing...images of women. It was then that she decided to devote her life to photographing women farmers and created the organization FarmHer.

FarmHer the Documentary, an Iowa Public Television production, is about Marji and the women she has profiled. The documentary follows seven women farmers from around the country and visits with them about what it's like to be a woman in what is typically thought of as a man's world. The FarmHers include:

- **Angelique**, a produce farmer from Iowa
- **Barbara**, a goat rancher from Texas
- **Carol Ann**, an urban farmer from Texas
- **Carrie**, a dairy farmer from Wisconsin
- **Pam**, a row-crop farmer from Iowa
- **Reneé**, a cattle rancher and exporter from Florida
- **Sara**, a sheep rancher from Montana

These inspirational women all share a deep passion for their chosen lifestyles.

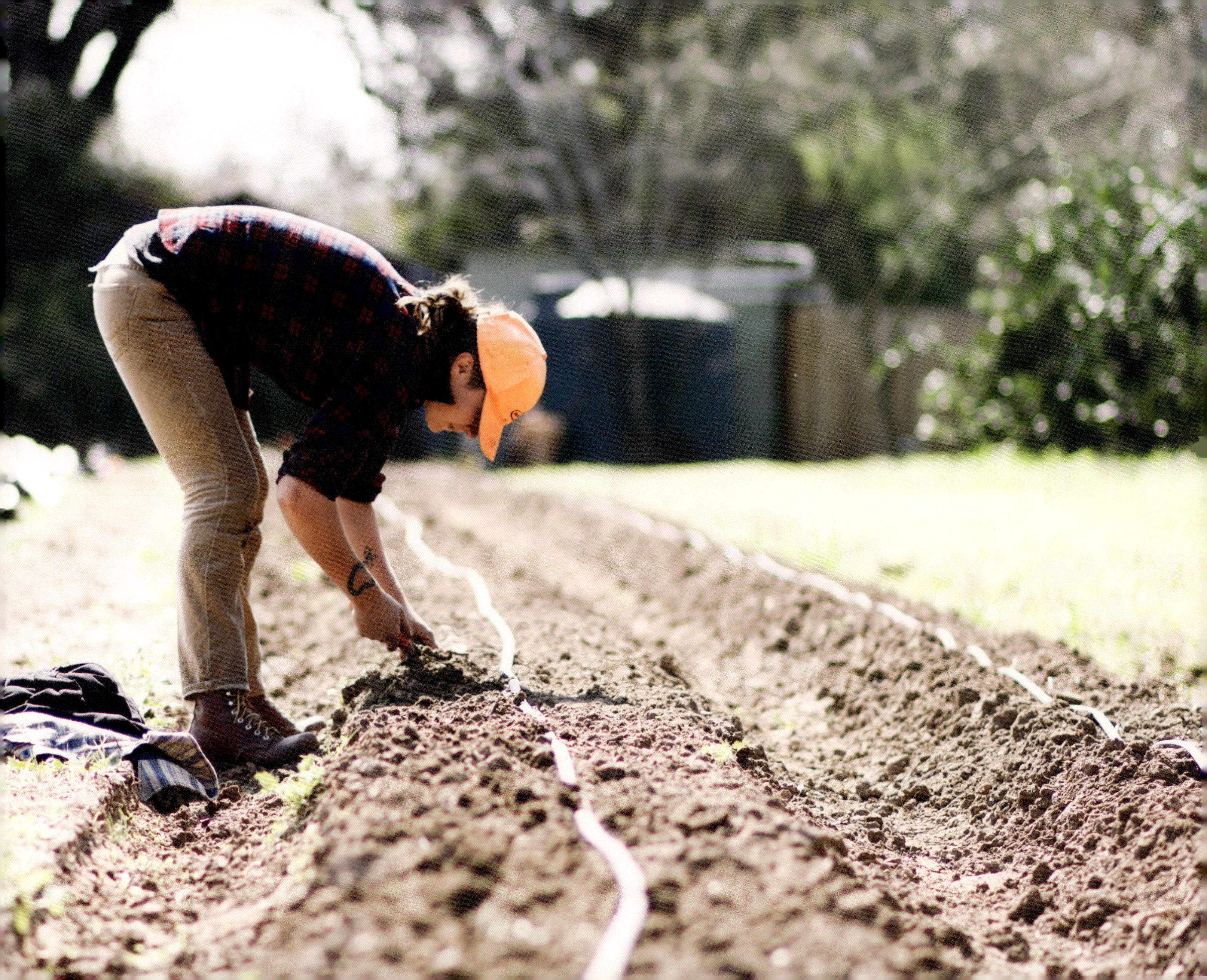

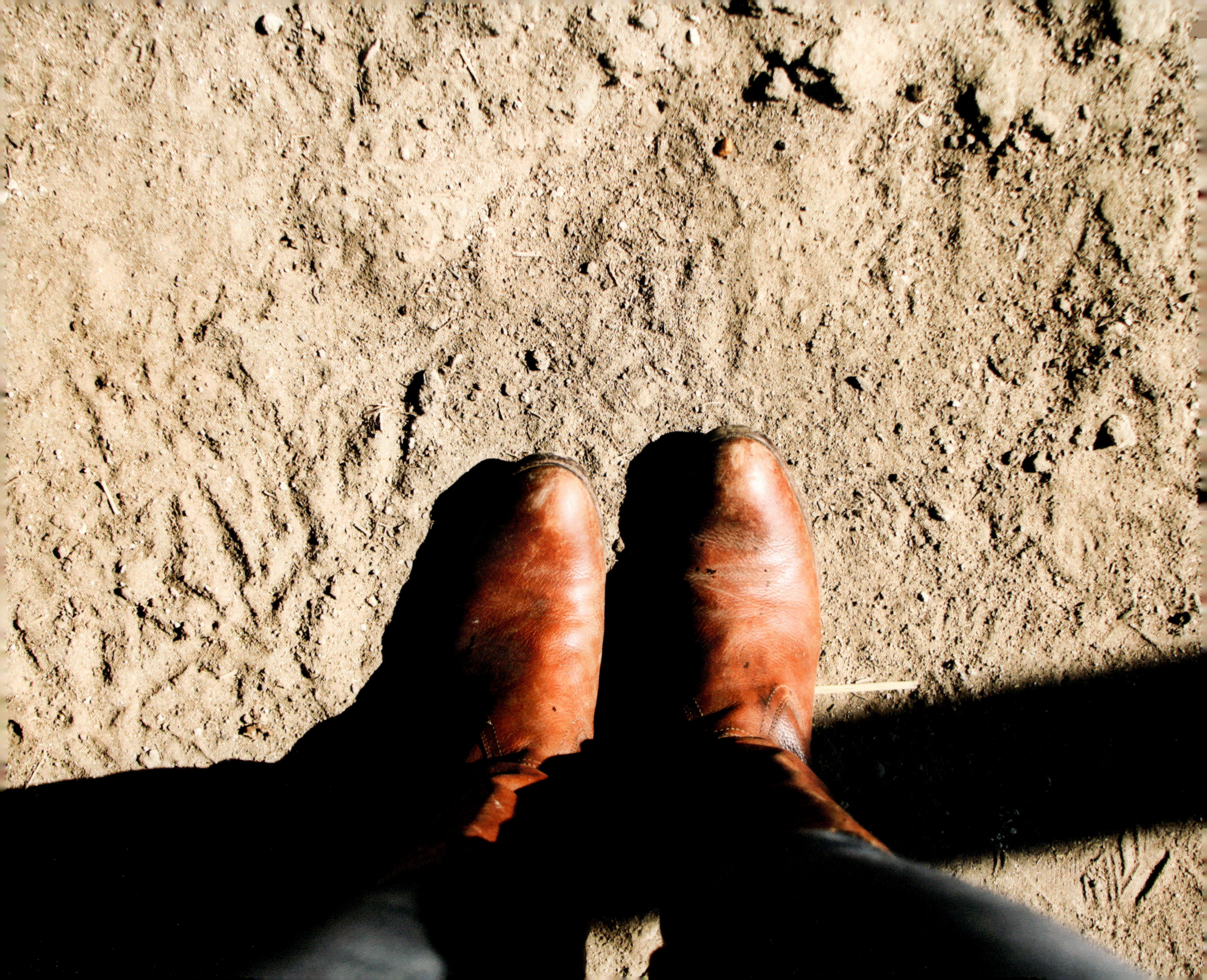

MARJI
PRESIDENT & FOUNDER • FARMHER

Marji Guyler-Alaniz, president and founder of FarmHer, is a lifetime Iowan and lover of photography. That love combined with graphic journalism and photography degrees from Grand View University, an 11-year career in corporate agriculture and an MBA from Drake University led her to launch FarmHer in the spring of 2013.

Guyler-Alaniz's goal with FarmHer is to update the image of agriculture by showing the female side of farming, creating community amongst women in agriculture and reaching out to young women interested in agriculture. Her work for FarmHer has been featured in a variety of publications ranging from *Smithsonian Magazine* and *Fast Company* to *O the Oprah Magazine*.

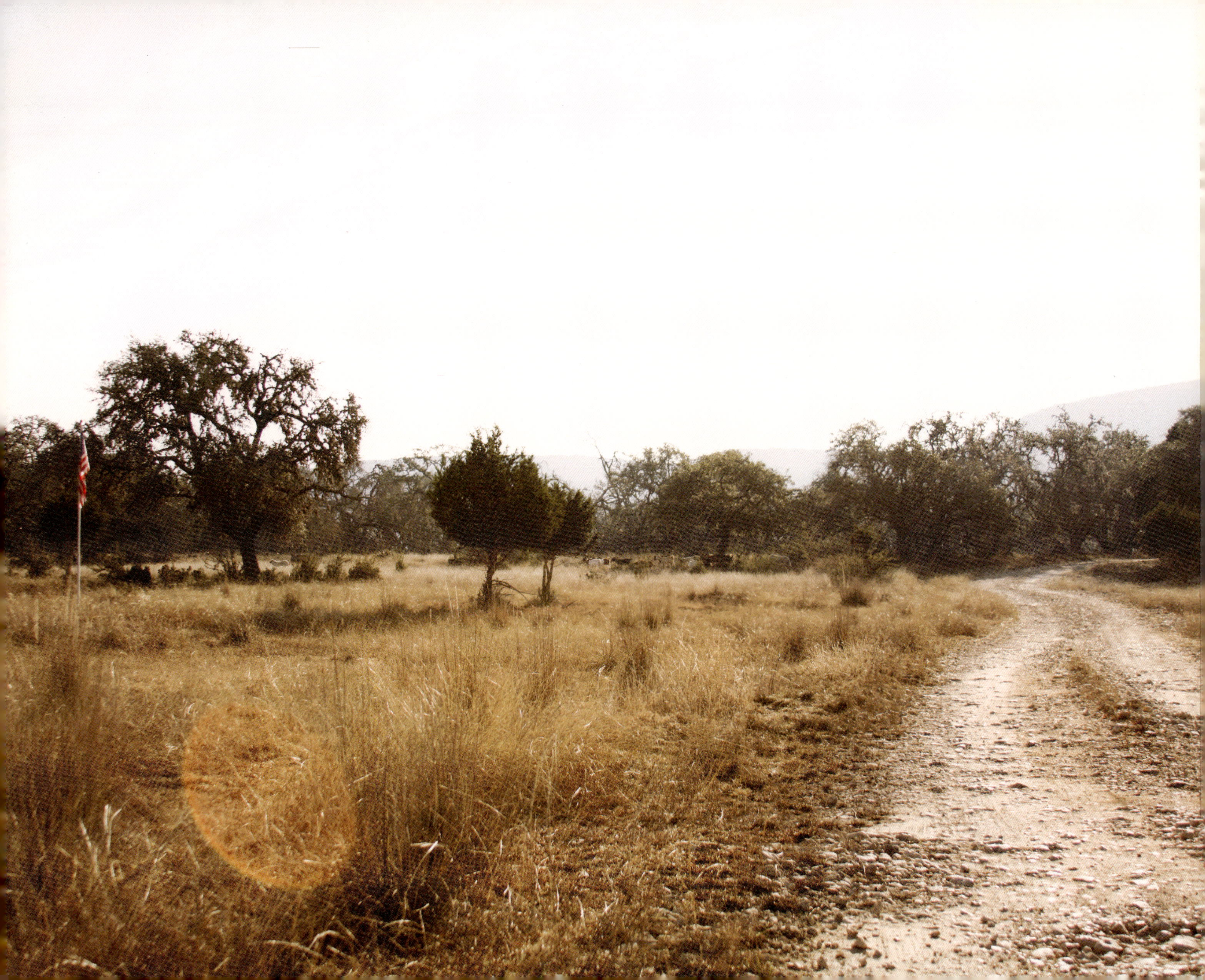